SMALL & UNUSUAL WOODTURNING PROJECTS

James A. Jacobson

Sterling Publishing Co., Inc. New York

For Peter, who said we should get a lathe. . . .

Acknowledgments

I wish to thank Mr. Russ Zimmerman of Putney, Vermont, for granting me permission to use his tool-sharpening jig design, which originally appeared in *The Zimmerman Turning Letters*. I also wish to thank Dremel, Division of Emerson Electric Co. of Racine, Wisconsin; Stanley Tools, Division of the Stanley Works of New Britain, Connecticut; and Conover Woodcraft Specialties, Inc. of Parkman, Ohio, for providing particular photographs.

I'd like to extend special thanks to the Honorable A. Andreas Matoesian of Granite City, Illinois, for contributing various projects and sharing his turning expertise. Thanks are also extended to Mr. Christopher Weggemann of Lawrence, Kansas, for sharing two of his favorite projects.

I'm especially grateful to my son, Peter J. Jacobson, for taking the photographs for this book in addition to contributing one of his turning projects.

EDITED BY LAUREL ORNITZ

Library of Congress Cataloging-in-Publication Data

Jacobson, James A.
Small & unusual woodturning projects.

Includes.
1. Turning. I. Title. II. Title: Small and unusual woodturning projects.
TT201.J33 1987 674'.88 87-9970
ISBN 0-8069-6510-X (pbk.)

Published by Sterling Publishing Co., Inc.
Two Park Avenue, New York, N.Y. 10016
Distributed in Canada by Oak Tree Press Ltd.
% Canadian Manda Group, P.O. Box 920, Station U
Toronto, Ontario, Canada M8Z 5P9
Distributed in the United Kingdom by Blandford Press
Link House, West Street, Poole, Dorset BH15 1LL, England
Distributed in Australia by Capricorn Ltd.
P.O. Box 665, Lane Cove, NSW 2066
Manufactured in the United States of America

CONTENTS

FOREWORD

This book presents a series of projects for both the beginning and experienced hobbyist woodturner. Since any number of books are available on woodturning, I've deliberately avoided any lengthy discussions on the basics of the craft; however, I do discuss various methods, tools, and techniques as they pertain to particular projects. I provide sufficient detail for the completion of each project, while still strongly encouraging your own innovation. Fortunately, by its very nature, the lathe permits this kind of creativity. Thanks to the lathe, our pieces rarely turn out exactly as we've planned them.

Unlike other books that present a series of projects designed by the author, this book takes a slightly different approach. Along with some of my own projects, I've included various projects developed by three hobbyist woodturners because I believe that some of the most exciting, innovative, and skilled turning is being done by amateur woodturners.

By definition, hobbyist woodturners are people who are employed full-time in some trade, business, or profession other than woodturning; they could also be full-time students. Working with the wood lathe has, for various reasons, become a means of relaxation and satisfaction for these individuals. The Honorable A. Andreas Matoesian, Circuit Judge of the State of Illinois, is a hobbyist woodturner who has consented to share some of his favorite turning projects. In addition to Judge Matoesian, my son, Peter J. Jacobson, who is a student at Washington University in St. Louis, Missouri, shares one of his favorite projects. Also, another hobbyist turner, Chris Weggemann, an architect who resides in Lawrence, Kansas, provides two box designs that he especially enjoys turning.

Judge Matoesian is a confirmed cutter who views scraping tools as primitive instruments of warfare. I, on the other hand, think scrapers can be very effective for certain projects and I tend to use them frequently. In addition to making the various projects, I hope you will also enjoy the occasional banter between friends over how best to turn wood.

INTRODUCTION

Unlike other large woodworking tools, the lathe quickly makes a compulsive captive of its user. Once you begin to work with this remarkable tool, you quickly discover that you can't leave it alone. The more you turn, the more you want to turn.

Unlike other tools, the lathe seems to contribute early on towards the success of the user. While an initial piece may come out smaller and shaped differently than planned, it will undoubtedly be unique and bring immense satisfaction. I suspect that most experienced turners still have their first turned piece and can remember the feeling of pride that accompanied its completion.

The lathe teaches us how to enjoy ourselves. It's a magnificent tool for the professional, but the lathe is at its best with the amateur turner. In an age cluttered with activities that tend to make all of us spectators, the lathe insists on total involvement and attentiveness. It demands that its user be an active participant with a singular focus. However, once it has engaged our attention, it encourages us to lighten up and enjoy ourselves. If you have a high-stress job, working on the lathe can be a relaxing way to begin or end the day. Or if you're faced with job "burnout," the lathe can provide an activity with a visible and satisfying outcome. In short, working on the lathe can offer a respite from our daily activities, which can make us feel good about ourselves and help us put our lives in perspective.

Working on the lathe can also provide the user with new experiences in terms of time. In contrast to the working world where time is often an unrelenting taskmaster, the very nature of turning makes us oblivious to clocks and the demands of schedules. Because the lathe has a way of engaging us so completely, once we've starting turning wood, we don't want to stop. We suddenly discover we've become a captive to this incredible tool and the experiences it provides.

Other shop tools are certainly very effective, but the wood lathe clearly stands alone in its ability to engage and hold us. For example, I enjoy using a table or radial arm saw, and my planer and jointer are marvels at what they can do with wood. These and other large shop tools are very efficient in performing their specific functions and, together, make woodworking possible. However, while you can develop new methods or clever jigs to use with these tools, they do not in themselves captivate those who use them. On occasion, my band and scroll saws attract my interest because of their inherent capacity for detailed and imaginative cutting, but in time I tend to tire with the persistent repetitiveness that accompanies the use of these tools.

However, to prepare the stock for the projects in this book, you will find that a band saw is indispensable. Many of the projects are made from pieces of scrap that frequently need resawing to make them usable. A number of projects also require other large and small shop tools. Often the best way to enhance a piece or make it more functional is to use a tool other than the lathe. Incidentally, you can scrape the revolving wood or cut it with a gouge—the choice is up to you.

Many of the projects are made from hardwoods that are indigenous to this country. However, I've also turned a number of them from unusual or imported woods as well as from that marvel of nature, the spalts. Since the projects are so small, many can be made from the scraps of wood that most woodturners have lying around their shops.

While many turners prefer on-lathe finishing, my own preference is to finish a piece off the lathe. In the final chapter, I discuss both methods as well as the finishing products and procedures that I've found useful.

Although each project includes specific ways of carrying out certain tasks, I encourage you to take advantage of your own skills and to make any necessary modifications in terms of design, techniques, and tools. The most important thing is for you to enjoy the process of working on the lathe.

I
BEAN-POT BOX WITH LID

As you might guess, this project involves faceplate turning. However, the project is large enough so that you can explore various turning techniques in addition to different tools. For example, while you may want to use a bowl gouge for shaping the base, I generally use a 1″ or ½″ roundnose scraper. In addition to these initial shaping tools, you may want to use a square-nose scraper to make the lid shoulder in the base. As you move through the various tasks in this project, you will make other decisions regarding tools and methods. Part of the joy of working with the lathe is that you can use several different approaches to achieve basically the same result.

Illus.1. Bean-pot box with lid

I turned the bean-pot base in Illus. 1 from black walnut that was 2½″ (10/4's) thick. The finished base has a diameter of approximately 4″. Given the cost of the thicker hardwoods, you may prefer using thinner stock. A good way to achieve the desired thickness is by gluing up thinner stock—more about this procedure shortly. The lid on the turned base in Illus. 1 is made from 1″ (4/4's)-thick stock. The lid knob is also made from 1″ (4/4's)-thick material. Illus. 2 is a dimensional drawing of the project, showing both the base and the lid. You can use this drawing as a reference for the overall design of the project.

TASKS:

1. If 2½″ (10/4's)-thick stock is not available, you may want to glue together three pieces of 1″ (4/4's)-thick stock for the turning block for the base. Assuming the surfaces of the boards have been planed, the final thickness of the glued-up block should be about 2½″. This certainly will be adequate for the project. Since the finished base diameter is approximately 4″, either your solid block or your glued-up block should be at least 5″ square. This dimension will give you sufficient wood for the base with a minimum of waste. You may, of course, prefer to work with a block that has a larger diameter and thus increase the diameter of the finished base.

An alternative to gluing together three 1″ (4/4's)-thick pieces from the same type of wood is using a different wood for the middle piece. For example, a contrasting piece of cherry, oak, or one of the exotics such as padouk would look very good with black walnut. You can also emphasize the oak, cherry, or exotic and use black walnut for the middle board instead. If you use contrasting woods for the base block, you may want to use a contrasting wood for the lid knob. Designing your base block from contrasting woods makes the overall project more interesting. It certainly makes it less expensive, especially if you have scraps of the appropriate size.

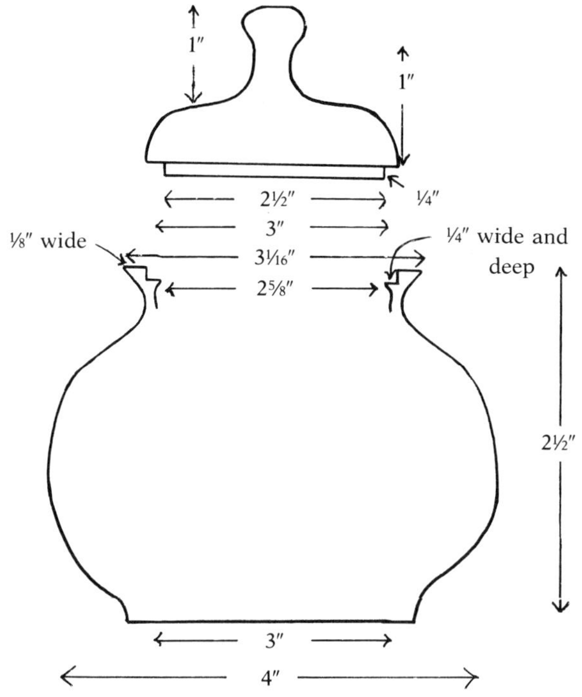

Illus.2. Base and lid

Another effective way to build the base block is to use veneers of contrasting colors or types of wood. For example, if you plan to use three pieces of 1″ (4/4's)-thick black walnut for the base, you may want to insert a piece of padouk or some other veneer between the wood sections. You can also double or triple the veneer thickness between two of the boards and use a single thickness between the other surfaces. When you are using veneer, remember to spread a sufficient amount of wood glue on the surfaces, use adequate clamps, and allow the assembly to properly dry. When you are turning with veneer inlays in the block, your tools should be sharp. Dull tools tend to tear out pieces of the veneer during the turning process. You need to use a light touch with your tools, especially if you are using scrapers.

Once you've solved your base-block problem, you should pattern and cut the block to an oversized diameter on the band saw. Given a finished diameter of 4″, the block should be cut oversized by at least ½″. With this diameter, you will have enough wood for turning the block down to dimension. (You may also want to turn the final base larger to minimize wood waste.) I generally use a school compass to make a circular cutting line on the block's surface. This little device is inexpensive and effective for patterning both base blocks and lids. For smaller projects, I use a plastic template with a range of circles of varying diameters.

For cutting turning blocks, I frequently use a ¼″-wide blade on the band saw. The type of blade or teeth per inch is not a major consideration when you are rough-cutting turning blocks. Often I will use my oldest blades for this operation. As a rule, I use ¼″ standard blades with six TPI (teeth per inch). These blades are effective for cutting turning blocks as well as for general purposes.

You also need to pattern and cut the lid block from 1″ (4/4's)-thick stock. Since the finished lid on the design has a diameter of approximately 3″, you can pattern and cut it to a rough diameter of 4″. This diameter should provide sufficient wood in case your base lid opening comes out larger than planned. If you want to be especially cautious, don't pattern and cut the lid block until you have completely turned the base. This way, you can actually measure the required lid diameter and cut the block accordingly. By following this procedure, you also can change the overall design of the base, if so desired. One of the joys and frustrations of working with the lathe is that our turnings do not always turn out as we had planned. This seems particularly true when dealing with exact dimensions.

For the lid knob, pattern a 1″-diameter piece

from 1″ (4/4's)-thick stock. If you prefer a longer knob, prepare it from thicker stock or reverse the grain direction of the knob block. This knob design is rather plain, but seems to go with the overall box-and-lid design. You may, however, prefer a knob that is a bit more elaborate or, at least, longer. If so, prepare the rough block accordingly.

After you have cut the lid and knob blocks to diameter, you need to glue them together. Select the best surface of the lid block for the top. After spreading wood glue on the middle surface of the lid and on one end of the knob block, assemble the two pieces and clamp them together. I use a bar clamp for this procedure (Illus. 3). When you are clamping, make sure that the pressure is evenly distributed on the top; this way, the entire surface of the knob will be flush against the lid's surface. If the knob slides on the glue during the clamping procedure, you need to recenter the knob and realign the clamp. The knob block should be centered on the lid's surface or it will be off-center when you begin turning. This usually results in a knob that is too thin or one that breaks off during the turning process. Allow the assembly to remain clamped until the glue is dry.

2. As with the other faceplate projects, there are a number of ways to secure the base block for turning. For those who have one of the commercial chucks, simply follow the procedures for your particular chuck. I do not have one of these chucks; so I use any number of alternatives for mounting the base block on the faceplate.

If you want to be certain that the block is well secured during turning, direct attachment of a 3″ faceplate to the block is a good option. You can enlarge the screw holes in the base with a ⅜″ bit later on, and then cut ⅜″ plugs and glue them into the holes. This procedure creates an attractive bottom, and is an excellent way to deal with screw holes. The plugs are sanded flush after the glue is dry.

Over the years, I've tried a range of different screws for securing faceplates to turning blocks. I always use an electric, variable-speed, reversible drill with a power bit secured in the chuck. Recently I've been using square-recess panhead screws with great success. The square recess assures an excellent drive and retrieval of the screws. I've also found that the square-recess panhead screws and the power bit for them hold up considerably longer than the Phillips devices I used to use. You can also purchase square-recess hand drivers if you would rather not use an electric drill. The type and thickness of your faceplate largely determine the length and type of head of the screws you use. On larger blocks, I like the screws to penetrate at least ½″ (Illus. 4). If you want the bot-

Illus.3. Lid-and-knob block: glued and clamped

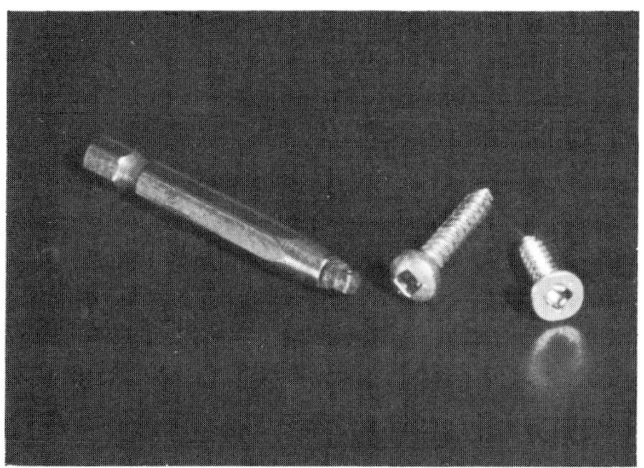

Illus.4. Square-recess screws and power bit

tom of your project to be a certain thickness, select screws that won't penetrate all the way. Square-head screws and drivers are available from any number of mail-order suppliers in addition to some of the better hardware stores. They're great—you may want to try them.

If I don't want screw holes and plugs in the base of a particular project, I use a pine or plywood disc along with white glue to secure the block. You should cut the pine or plywood disc to a diameter that is at least ½″ larger than the faceplate. This helps prevent the disc from splitting when you drive the faceplate screws into it. Plywood is less likely to split than the solid disc. I tend to use regular 1″ (¾″)-thick pine for the glue discs. When using a 3″ faceplate, I pattern the discs with a 3½″-diameter circle from a plastic template and then cut them on the band saw.

Although you can use white glue and paper between the glue disc and the turning block, I usually omit the paper. When paper is used, it's easier to split the glue disc from the finished block, but it really leaves a mess on the bottom of the turned piece. You can sand off the glue and paper using a large belt sander and a coarse belt. Unfortunately the glue-and-paper combination tends to gum up the abrasive belt. When I do use this type of procedure, I often separate the disc and the turned piece with an old hand-plane blade and a mallet. Be careful that you don't gouge the bottom surface of the turned piece when you are using the blade. You can also use a wood chisel to separate the two pieces.

If you plan to use the glue-and-paper method, be sure to spread the glue over the entire surface of the glue block and also on the top surface of the paper. The assembly must be clamped until the glue is dry. Be certain you center the glue disc on the base of the turning block. Also, when clamping, be certain the disc doesn't slide off-center when you are applying pressure on the clamp. I use bar clamps for this procedure. When the glue is dry, mount the faceplate to the glue disc.

Another method you may want to try involves gluing the disc directly to the bottom of the turning block. I prefer this method because, when I've completed the turning, I can saw the glue disc from the turned piece using the band saw. Simply spread white glue on the precut disc and place the disc on the bottom center of the turning block. Clamp the assembly using a bar clamp. On blocks of sizable diameter, you may have to use two clamps and a board. It's a bit awkward to clamp an assembly this way, but it does work. Be sure you allow the glue ample time to dry. Secure the faceplate to the glue block and turn.

You must be careful when you are cutting the glue block off the bottom of the turned piece. Be sure to support the turned piece on the table of the band saw. I leave the faceplate attached because it gives me something to hold onto during the sawing process. It also helps guide the assembly into the blade. If the assembly is not supported and if you don't have a tight hold on it, the band-saw blade can pull it from your hands. The initial penetration of the blade is usually the point when problems will occur. Take your time with this procedure. I cut slightly into the bottom of the turned piece and thus totally eliminate any problems with the glue. This is an effective way to remove the glue disc without the usual mess.

There are also some options for turning the lid and its knob. I usually secure the lid block to a pine disc with double-face tape. You should secure the pine disc to the faceplate and then attach strips of tape to the disc's surface. After removing the protective layer of the tape, center and place the lid block on the tape. Clamp the assembly for a few minutes using a bar clamp or another clamp that is similar. You want a good bond between the tape and the block. As a rule, with larger blocks, I place three strips of the tape, side by side, on the surface of the pine disc.

You definitely should use the commercial double-face tape that is available at various turning-supply outlets. The kind of double-face tape found at local hardware dealers and used with carpeting will not work. It simply does not offer a strong enough bond to hold the block during turning. Although the commercial tape is rather expensive, it's simple to use and very effective. Incidentally, the tape should not be used a second time. Sometimes it holds, but it's unpredictable. You're better off simply using new strips of tape for a second project.

If you've never used the tape, you may need to develop some confidence in its use. When turning the lid and its knob for this project, you can always put a revolving center in the tailstock and secure it against the knob. This is always a good safety precaution, but it's especially helpful when you are using tape for the first time (Illus. 5). My own experience with the tape has been very successful. Whether scraping or using a gouge, the tape holds the block securely in place. You will discover that you will need to pry the finished turning off the tape. When you are removing the newly turned piece, be careful that you don't damage its edges. I generally use an old hand-plane blade or a large wooden chisel to force the turned piece from the tape. When you apply steady pressure under the edge of the piece, the tape will begin to stretch and eventually release the piece.

You can also directly glue the lid-and-knob assembly to a pine or plywood disc. The procedures are the same as those described earlier for securing the base block. Be certain the lid block is centered on the disc when you are gluing and clamping the assembly. After completing the turning, you can separate the finished lid from the glue disc using the band saw. It's a good idea to leave the faceplate attached when you are sawing the lid piece from the disc. It gives you something to hold onto and you can also use it to guide the assembly into the blade. You may also want to support the entire assembly on the band-saw table prior to cutting. Be very careful not to damage the tapered edge of the lid during these procedures.

Another way to turn lids with knobs involves using a screw chuck. Although any number of commercial screw chucks are available, my budget for turning accessories always seems to be nonexistent; so I usually resort to making my own chucks. They are not very attractive, but they're inexpensive to make and very effective.

Actually, there are two different types of screw chucks that I make and use for turning box lids. You can use both of these chucks for other purposes, as later projects will indicate. Since one of the screw chucks is made from 1″

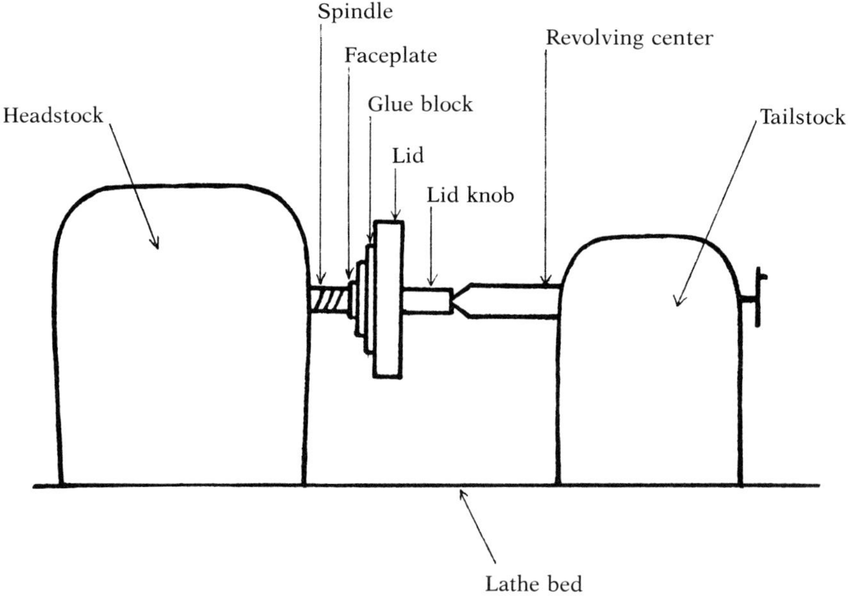

Illus.5. Setup for turning lid and knob

(4/4's)-thick stock and the other from 2″ (8/4's)-thick stock, you may want to make the one that best meets your general needs. Or, if you don't have 2″ (8/4's)-thick stock on hand, you may want to make do with the thinner one. Both chucks, mounted on 3″ faceplates, are shown in Illus. 6. If you can afford it, buy some extra 3″ faceplates or have some made by a local machinist. It's a good idea to leave the faceplates attached to the chucks for future use.

I usually make the 1″ (4/4's)-thick screw chuck from hard maple. Unlike the red oak or ash that I use for the 2″ (8/4's)-thick chuck, the maple rarely splits when penetrated by the faceplate screws. You may want to drill small pilot holes into the block to avoid any possible splitting. In addition to the wood, you also need ¼″-diameter lag screws that are at least 2½″ long. As indicated, the chucks are attached to 3″ faceplates.

I pattern and turn the chucks to a final diameter of about 3½″. Not only does this allow sufficient wood for the chuck, but it also reduces the chance of the wood splitting when penetrated by the faceplate screws. Pattern and cut your blocks to a diameter of at least 4″. Mount and secure the faceplate to the block. If you are using 1″ (4/4's)-thick stock, select screws that won't penetrate the block. You don't want screws sticking through the face surface of the chuck. Be sure, however, that the screws are long enough to hold the entire assembly on the faceplate when you are turning. You don't want the assembly to fly off the faceplate during the turning process.

Using a ½″ or 1″ roundnose scraper, turn the blocks to the preferred diameter. On the 1″ (4/4's)-thick block, you want to make a tenon that is ¼″ long and has a diameter of about 2″. This tenon on the front surface of the chuck makes it easier to remove the turned piece from the lag screw. Without a tenon, the piece tends to lock against the chuck's surface during turning. If you are making a 2″ (8/4's)-thick chuck, you also will want to make a tenon. Since I use this larger chuck for a range of projects, I turn the tenon to a diameter of 2¼″ and a length of 1¼″. You may want to refer to Illus. 6 to note the basic differences between the two chucks.

After you have turned the tenons, realign the tool rest to the front of the chucks. Using either a skew chisel or a square-nose scraper, clean and square the front surfaces. Also, cut a dimple in the middle of the front surfaces by holding the point of the skew or the corner of the scraper on the middle surface of the chuck. You need to drill the lag-screw hole through the chuck at this point.

Illus.6. Homemade screw chucks

Since a ¼″-diameter screw can't penetrate the middle of most faceplates, you need to remove the faceplate from the chuck. Before removing the faceplate, make clear alignment marks on the edge of the faceplate and the back surface of the chuck. These marks will permit you to put the faceplate back in the exact position it was in before. After you have removed the faceplate, you can proceed with making the chuck.

First, on the back surface of the chuck, find the approximate center and drill a ¾″-diameter hole about ⅜″ deep. This hole will not only accommodate the head of the lag screw, but also the socket from a socket wrench, which you will need to use to screw the lag screw through the chuck.

Next, using a 3/16″ (or smaller)-diameter high-speed bit, drill a hole through the chuck, beginning at the dimple on the front surface. You want the lag-screw hole undersized so that the lag screw will not unscrew during use. If it does loosen a bit, the faceplate will stop it from totally unscrewing. The best way to do the drilling procedures is by using a drill press.

I place the chuck in a wood vise in order to screw the lag screw through the chuck. The vise holds the chuck in place while you screw in the lag screw using a small socket wrench. You want to screw the lag all the way through the chuck so that its head is snug against the surface. After the lag screw is in place, reattach the faceplate according to the alignment marks. If you don't properly reattach the faceplate using the alignment marks, the chuck will be off-center.

Mount the chuck on the lathe and, with a hacksaw, cut the extending lag screw to a length of 3⁄4" from the front surface. This length is usually adequate for most projects that require the use of a chuck. On a project where the screw is too long, I thread a shim between the chuck surface and the piece to be turned.

With the chuck secured on the spindle, turn on the lathe and watch the lag screw as it rotates. If the screw is not running true, shut off the lathe and tap the screw in the appropriate direction. A bit of trial and error usually results in a true-running screw. So that the screw will have more holding power, I deepen the gullets of the lag with a small file. You can accomplish this by rotating the chuck by hand while it's mounted on the lathe. The deeper gullets tend to grab more wood on the block being turned, allowing for a safer assembly.

To use the chuck—when you are turning a box lid, for example—you need to drill an undersized hole into the middle bottom of the block. Usually a 3⁄16"-diameter high-speed drill bit will do the job. You want a small enough hole so that the 1⁄4"-diameter lag screw can firmly hold the block in place. You may need to experiment with various woods to obtain the best diameter. Once you have drilled the hole into the block to be turned, thread it onto the lag screw, flush against the chuck's front surface. For additional security and stability, I frequently force a revolving center from the tailstock into the piece. The locked tailstock and revolving center will securely hold the chucked block in place (Illus. 5).

3. After attaching the faceplate to the base block for the box, align the tool rest according to your own preference. This alignment partially depends upon the type of turning tool you plan to use. As a rule, I set my tool rest a bit above the middle line of the block when I'm using scrapers. When I use a large roundnose scraper to true the block, I leave enough room between the tool rest and the block so that I can point the scraper slightly downwards. I want enough space to move the scraper but not so much that I lose the support of the tool rest. When I'm using a gouge, I move the tool rest out and down a bit so that the bevel can easily ride the block. I find myself raising the tool rest and moving it closer to the block when I'm using a skew chisel to plane the surface. Actually, I've found that how and where I place the tool rest may have more to do with habit and comfort than with any underlying rationale; this may also be your experience.

4. As previously indicated, I frequently use a 1" roundnose scraper to rough out the base block for this project. I true up the wall over the full length of the block. Before turning the base block, you may want to place a revolving center in the tailstock and bring it forward for support. Although the base block for the project is not very large, the center and tailstock do give you a sense of security when you are doing faceplate work. For very large blocks, I always use the revolving center and the tailstock for support when I'm turning both the outside and inside of a piece.

I admit that I never sharpen my turning tools after a stint at the lathe. Although I sharpen or hone them frequently while turning, I tend to leave them dull when the lathe work is done. In any event, I sharpen my scrapers, chisels, and gouges on an electric grinder using a medium-grade Carborundum (silicon-carbide) wheel. The wheel is 1" wide and has a 6" diameter. I'm waiting for it to wear down so that I can buy one of the friable aluminum-oxide wheels, which are more effective and generate less heat in the tools when they are grinding. I periodically clean my present wheel using a dressing stick to prevent glaze. On rare occasions, I'll reshape the stone using a dressing wheel.

Incidentally, I rarely dip the tools in water when I'm sharpening them. This is mainly because the small plastic glass I keep next to the grinder is usually empty. If I think of it, I'll use water when grinding a carbon-steel tool, but with high-speed steel, I never dip the tool in water when I'm sharpening it. It's curious how the experts seem to disagree when it comes to whether or not to dip tools when sharpening them.

I've found that if you minimize the pressure on the tool and keep it moving while touching the stone, you can avoid a lot of overheating. And, as I've said, the friable wheels also help reduce heat build-up. In the next chapter, I describe an effective device to use when sharpening gouges.

I sharpen my scrapers while I am holding the tang against a small tool rest on the grinder. The tool rest not only serves as a support, but also as a guide for my bevel. I usually grind about a 15° bevel on my scrapers. This angle tends to make the edge last a bit longer. As you may know, you should leave the wire edge that develops during the grinding process. It is the edge that makes the scraper work. To keep the scrapers effective, you need to grind them often. Tool steel is relatively inexpensive, and regular grinding assures you of pieces with better surfaces. Regular grinding also helps reduce the need for extensive sanding.

Before proceeding further, a word about turning tools in general seems to be in order. I tend to use tools according to what works best for me on a given task or project. Since much of my turning has been production turning of boxes, I've come to rely heavily on the scraping method, but I always try to use a ½″ or 1″ skew chisel (hollow-ground) to plane the surfaces. When turning for my own satisfaction, I will frequently spindle-turn using a 1¼″ roughing gouge or ¼″, ⅜″, and ½″ fingernail gouges ground a little more blunt than usual. As with many turners, I continue to develop new skills with the cutting, or shearing, method.

As you will discover in Chapter 2, Andy Matoesian is a hard-core cutter. Andy is of the Frank Pain school that believes that you should only cut wood the way it prefers to be cut—and it prefers to be cut, not scraped. My usual response to this dictum is that only a fanatical cutter would talk to wood and believe it talks back. But in Chapter 3, Andy condescends to demonstrate a project using scrapers. His skill with scrapers shows that he, like all cutters, uses scrapers but doesn't like to admit it.

You can use either the scraping method or the cutting method with most of the projects in this book. Or you can use both turning methods on a particular project. It's really up to you. If the tools do the job and you're enjoying what you're doing, by all means use them. The emphasis on the right way and the wrong way can take much of the fun out of turning. Now, back to the project.

Generally I run my lathe at 2220 rpm when I'm doing faceplate work. After securing the base block to the lathe, I true it using a 1″ roundnose scraper. You may prefer using a roughing-out or bowl gouge for this procedure. After the block has been rounded, I do some initial shaping with a ½″ roundnose scraper. You may want to refer back to Illus. 2 to note the box-base configuration again.

Before making the wide cove area in the neck of the base, I reduce a ¼″-wide section of the block's front edge. Using a ½″ (or larger) square-nose scraper, I reduce this area down to a diameter of 3½″. This diameter is the approximate final dimension of the top of the box where the lid will rest. The procedure makes cutting the neck cove easier and assures a fixed diameter for the lid area. You can easily make the neck-cove area using a ½″ roundnose scraper or a ¼″ fingernail gouge. If you use the gouge, be sure to keep the bottom of the cove open. As you can see in Illus. 2, the cove begins at the top edge of the base and moves down the side. This tends to give the cove a long, sweeping appearance. Using the same tool, make a partial cove on the bottom edge of the base. Before I apply any abrasive papers to the base (if they're needed), I turn the inside area. Only after the entire base is turned do I use the abrasives.

5. Realign the tool rest so that it faces the front of the block. If you are using the tailstock as a support, you can leave it in place. Use a 4″-long tool rest rather than the standard tool rest that is 12″ long so that you will have enough room to turn, even with the tailstock in place. Before cutting the lid shoulder and removing wood from the inside of the base, you may want to refer back to Illus. 2.

Before cutting the lid area, begin to remove wood from the inside of the base. You can easily accomplish this using a ½″ or ¾″ square-nose scraper and a 1″ roundnose scraper.

With the square-nose scraper, begin removing the wood just left of the center of the base. Leave a ½″-wide area around the top edge for making the lid recess. You can cut this area to dimension after you've removed some wood. I generally go right straight in with the scraper. Move the scraper to cut out different portions of the base, and be careful that you don't overheat it. Give it a breather. You will see that a pillar will develop in the middle under the revolving center. This is the support pillar, which you should leave in place as long as possible.

After removing some wood from the inside area, you will want to cut the lid area. I don't shape the inside wall until I've turned the lid-recess area. When you do shape the inside wall, it's best to use a 1″ or ½″ roundnose scraper. If you've used a glue block instead of screws, you can have a bottom thickness of ¼″ or less. Periodically check the depth of your cuts so that you don't go through the bottom of the block. You can make the final cleanup cuts on the bottom. Remove the support pillar after you've removed the wood from the wall.

When most of the wood has been removed from the inside, cut the lid-recess area. But before you begin, regrind the square-nose scraper, which should have a good wire edge so that it will make clean, sharp cuts. A dull scraper tends to chip the edges. If you don't have a square-nose scraper, use a ½″ square-nose chisel, but use it as you would use a scraper. The lid recess should be made ⅛″ in from the outer wall. In effect, you will have a ⅛″-wide wall that surrounds the shoulder where the lid rests. Using the scraper, cut the recess ⅛″ in from the outside wall and ¼″ deep. Obviously these dimensions are approximate, but you don't want the wall any thinner than ⅛″ or it may chip out during use. The lid shoulder should be approximately ¼″ wide. Remove any excess width from the shoulder area. When you have finished the lid recess and shoulder, you can remove any remaining wood from inside the base.

To round the inside wall, begin to remove some wood using a ½″ or 1″ roundnose scraper at the inside edge of the lid shoulder. Point the scraper towards the wall as you make the entry cut. You can then simply sweep down the wall with the scraper. I shape the inside of the wall to conform to the outside. Generally, on this design, I try to make the wall a uniform thickness of about ¼″. This tends to make the finished box feel much lighter than it appears. Of course, you may prefer to leave more wood and have thicker walls. When the inside is completely turned out, flatten the bottom surface with a square-nose scraper. You can also do this procedure with a ½″ square-nose chisel used as a scraper. It doesn't help the edge of the tool, but it certainly cleans up the bottom surface. The chisel is also a good tool for removing the inevitable protrusion that develops in the middle of the base. You may also want to clean up the shoulder and the lid-recess area with the square-nose chisel. Use a very light touch and be certain to use it as a scraper—not as a chisel. The sharp edge of the chisel will leave neat, clean edges in the lid area.

6. If your tools were reasonably well ground, you shouldn't need to do too much sanding. I generally begin with 100-grit, move on to 150-grit, and finish with 220-grit. With an uneven surface such as the base of this project, it's a good idea to use some backing with the abrasive paper. I generally cut pieces of carpet pad to use as backing. It helps in reducing friction and also allows you to sand in coves and other indentations. You don't have to wear a dust mask when turning, but you should definitely wear one when sanding. Even though you may not be sanding for very long, you will still generate a lot of dust from the process.

When using abrasive paper, it's a good practice to occasionally move it at an angle across the block's surface. This helps to cut off fibres that tend to lie down with straight sanding. When the cross grain is rough or mousey, I sometimes stop the lathe and spend a little time working on it. But don't rub too hard, and work with the contour of the piece. It also helps to have a reversing switch on your lathe, which allows you to sand off fibres that otherwise tend to lie down.

As you move across the surface with the abrasives, be careful not to cut the sharp edges of the lid recess. I use the 220-grit to clean up the sharp edges. It helps prepare them for finishing but does not significantly reduce them. You may prefer a rounded look on these edges; if so, sand them accordingly.

I generally wear a glove when I'm sanding the inside of the base. Since it's sometimes difficult to use backing with abrasive paper, the glove tends to absorb the friction and it also saves the fingers. Although I spend time working on the inside of the base, I actually view the inside as less critical than the outside. Before you use the glove, check to be certain there are no sharp extensions that could snag it.

You may want to apply a few additional finer grits to the surface, but this depends upon the type of finish you plan to use. On some pieces I will use 240-, 340-, and 400-grit abrasive paper. As with all finishing procedures, your choices depend upon how much time you want to spend on a given piece. Chapter 43 presents products and procedures that I've found useful in the finishing process. If you plan on doing on-lathe finishing, you may want to refer to this chapter before proceeding with the project. My own preference is to do all the finishing off the lathe.

7. Before starting to turn the lid, you may want to remove the turned box base from its faceplate or glue block. If you do not have a band saw to cut the glue block from the base, you can separate the two while on the lathe using a parting tool. I prefer the band-saw method, but the parting-tool approach works just as well.

To remove the glue block from the base, align the tool rest along the side of the turned base so that it extends beyond the bottom edge. You need the support of the tool rest while you are using the parting tool. The parting tool should be ground and honed so that it will cut the wood. Since I don't completely cut the glue block from the base with the parting tool, I let the lathe run at normal speed. You want to leave a connecting post of about ¼" in between the glue block and the base. After making the cut with the parting tool (leaving the connecting post in place), turn off the lathe. While holding onto the turned base, saw through the post using either a hacksaw or a handsaw. Finish the bottom surface using a belt sander with an appropriate grit. You also ought to sand the bottom with an electric pad sander or by hand, using the same abrasive grits that you used on the base wall.

If you prefer to band-saw the glue block from the base, leave the faceplate secured to the assembly. The faceplate gives you something to hang onto while you are making the cut. To keep the procedure safe and to prevent any possible damage to the turned piece, I place support scraps under both the turned piece and the glue block (Illus. 7). This procedure steadies the entire assembly and usually prevents the saw blade from pulling the assembly out of your hands. It's during the initial contact between the saw blade and the wood that you may have some problems. The support scraps under the turned piece and the glue block also help prevent possible surface damage to the piece from the band-saw table. You want to saw slightly into the bottom surface of the turned piece so that you cut off not only the glue block but also the mess on the bottom surface. Use a large belt sander to initially clean the blade marks off the bottom surface. Follow this procedure with finish sanding. You can use an electric pad sander or do the sanding by hand. Use the range of abrasive grits that you used on the box wall.

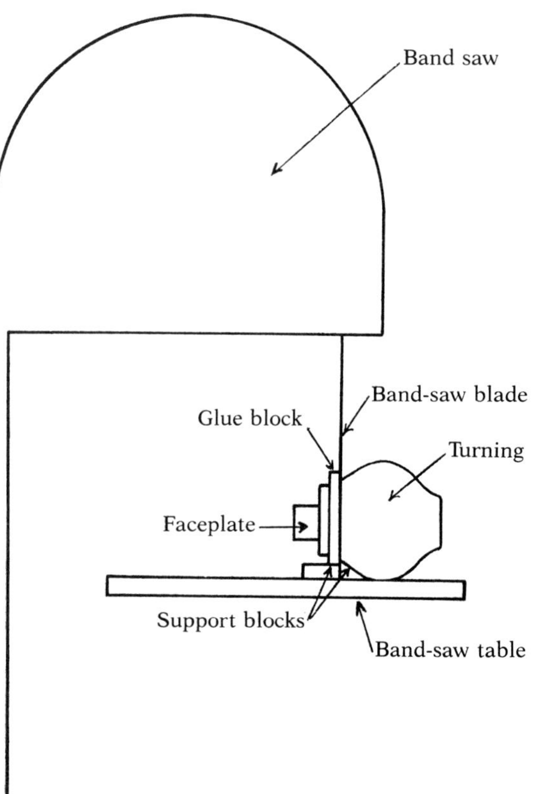

Illus.7. Separating turning from glue block with band saw

8. The next task in the project is to turn the lid and its knob. You may have already prepared the lid-and-knob assembly when you cut the

base block. If not, you should prepare it now according to earlier directions. For this task, you will want to measure the final diameter of the base lid area. Be sure to add at least ½″ to this diameter when you are patterning the rough lid block. You need this extra wood for turning the lid down to its final diameter. A good device for making this kind of measurement is a vernier caliper (Illus. 8). The caliper allows you to take both inside and outside measurements of a diameter. You will definitely want to have a caliper available when you are turning the lid. It is also a necessary device for a number of other projects in this book. You can buy a relatively inexpensive caliper that is more than adequate for your turning needs at a local hardware dealer or from a mail-order supplier. You do not need a precision instrument.

As indicated earlier, there are three basic ways of securing the lid block to a faceplate: You can use a glue block, double-face tape, or a homemade screw-chuck. If you haven't decided which procedure to use, you may want to review the earlier discussion on these methods. For the purpose of demonstrating the turning of the lid and its knob, I use a homemade screw chuck.

Since you will be securing the lid block onto a ¼″-diameter lag screw, you need to drill an undersized hole into the middle bottom of the block. The drilled hole should be as near the middle as you can make it. I generally "eyeball" the bottom of the lid block, approximate the middle in relation to the knob block, and then make a pencil mark for drilling. The hole must be drilled directly under the knob block and, if it's properly placed, partially into the knob. For a very long knob, after it has been turned, I often use this pilot hole for a wood screw or a piece of dowel to add support to the knob. You can drill another hole to accommodate the head of a screw and a wood plug to cover the head.

Using either a drill press or a small electric drill, make the pilot hole into the lid-block bottom. I use a 3/16″-diameter (or smaller) high-speed steel bit to make the hole. This diameter is small enough to afford good holding power on the ¼″ lag screw.

With the screw chuck on the lathe, thread the lid block onto the lag screw. The bottom of the block should be snug against the front surface of the chuck. Partially for safety reasons, place a revolving center in the tailstock and secure it against the top surface of the lid block. While

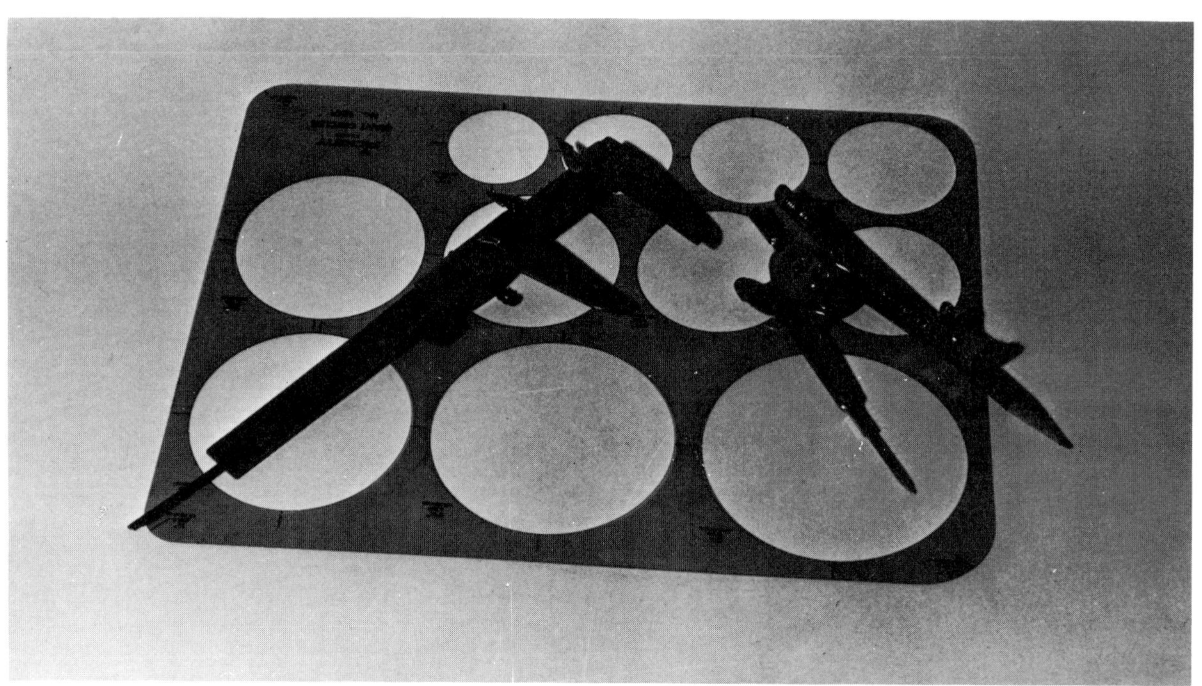

Illus.8. Vernier caliper, school compass, and plastic template

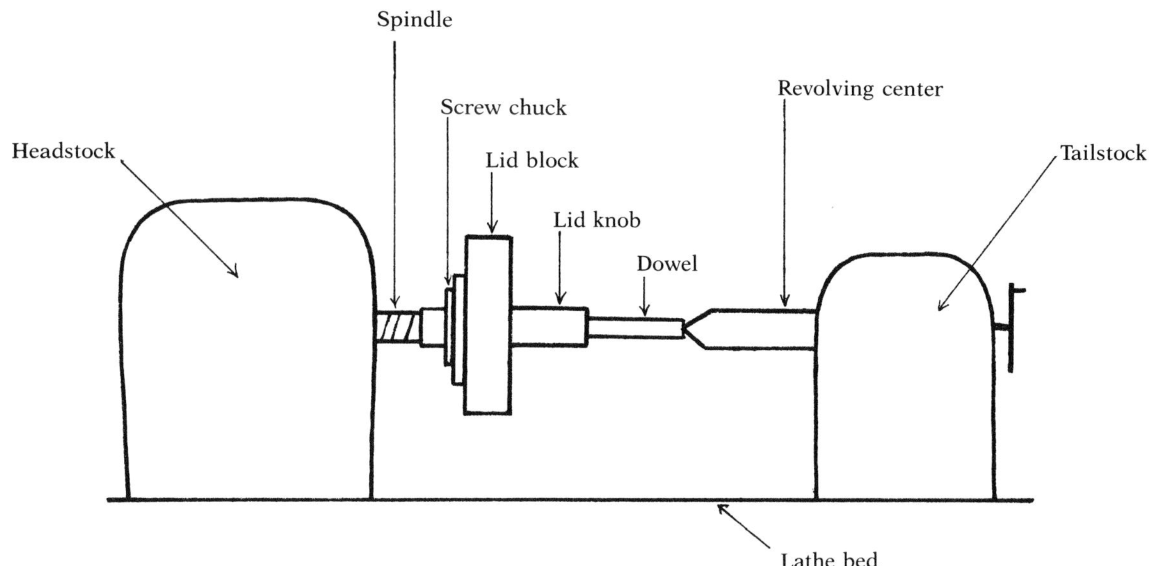

Illus.9. Setup for turning lid and knob with dowel-protecting knob

this holds the assembly on the screw chuck, it also offers needed support for the knob when it is turned. This procedure is almost mandatory when you are using a very long knob on the lid. Depending on how you plan to turn the knob, you may want to place a small block or dowel between the revolving center and the knob's surface. This prevents the revolving center from damaging the top surface of the knob (Illus. 9). You should remove the tailstock when you are doing the final turning on the top surface of the knob.

Before doing any shaping, you need to true both the lid block and the knob. I do this with a ½″ or 1″ roundnose scraper, but you may prefer to use a gouge. Using a vernier caliper, set the internal points on the inside area of the base lid recess. This procedure automatically sets the external arms of the caliper. As you true the lid block, monitor its diameter with the preset caliper. I tend to leave the lid diameter a bit oversized to allow for finishing cuts as well as any abrasive work that may be needed. You want the lid to fit neatly into the base recess, but it shouldn't be too snug or too loose. Using the caliper, plan your wood removal accordingly.

Prior to shaping the lid and knob, you may want to refer back to Illus. 1 to see the finished project. Although you may decide to duplicate this type of lid-and-knob design, there are many other options. I tend to do a lot of pencil-and-paper design work before actually turning a project. You may want to get in the habit of this kind of planning procedure. It really helps you think through the actual turning process and it also forces you to exploit your own creativity. Design the lid and knob the way you want them to look and then turn accordingly.

When I'm shaping the lid and knob, I use a ½″ roundnose scraper. Not only does this tool make possible the kind of shape I want, but it also permits me to clean up the glue line where the lid and knob join. Cleaning up the glue line isn't always easy. You want to align your tool rest so that you have good tool support and access to the area. Sometimes I will use a ¼″ roundnose scraper with a very light touch to make the final pass over the glue line. This procedure usually works and leaves an almost invisible glue line.

After I've shaped the lid surface and the knob, I remove the tailstock in order to turn the top surface of the knob. You need to realign the tool rest so that it's directly in front of the knob's top surface. Good tool support is critical when you make your cuts on this surface. Use a very light touch when shaping and cleaning up the knob's surface. Too much pressure with the tool can break off the knob at this point in the process. I use a ½″ roundnose scraper to turn the knob's top surface.

With the turning completed, apply abrasive

paper of varying grits to the lid and knob surfaces. When sanding the knob, be careful that you don't break it off inadvertently. Let the abrasive do its job without applying too much pressure. The longer the knob, the more caution you must take when you're sanding it. But should you happen to break off the knob, all is not lost. You can glue it back together. With a touch of wood glue on both surfaces, carefully align the break and clamp. Be careful when you are clamping that the pieces are not forced out of alignment. When the glue is dry, remount the assembly on the chuck and sand over the glue line. If the knob is a little off-center, replace the tailstock support system and turn the knob again. This procedure usually solves the problem and prevents you from having to turn a new lid. If the break is high on the knob, you can also turn the knob again and simply make it a bit shorter.

After you have removed the turned lid from the chuck, check it for fit in the box base. If it's too tight, reattach it to the chuck and reduce the diameter using abrasive paper. If your vernier calipers have been set correctly and you've checked the block with them while turning, you should have a good fit.

9. Depending on the type of wood you've used for the lid, you can fill the lag-screw hole with appropriately colored wood filler or plastic wood. Another option is to carefully drill a ⅜"-diameter hole over the lag-screw hole and glue in a wood plug or piece of dowel. You can make your own plugs using a ⅜"-diameter plug cutter. I often cut plugs for both the lid and the bottom of the base from scraps of exotic wood. The contrasting colors can enhance the overall appearance of the piece. And, since the holes are there, you might as well take advantage of them.

Using the appropriate abrasive grits, sand the lid's bottom surface for finishing. You may want to refer to Chapter 43 for a discussion of some finishing products and procedures.

2
BALL-POINT PENS

Ideally suited for woodworkers who enjoy spindle turning, this project is one of Andy Matoesian's favorites. Turning pens can be a challenge in terms of your skill and ability to use numerous lathe accessories. But once you've mastered the various procedures and techniques, you will no doubt, like Andy, want to make pens in quantity for friends and acquaintances.

Illus. 10 shows just a sampling of the pen designs Andy has turned. Not only does this project present a diversity of design possibilities, but it also offers a variety of options in terms of woods. Andy uses a wide selection of exotic woods in addition to those woods that are indigenous to the part of Illinois where we live. He has a special fondness for turning pens and

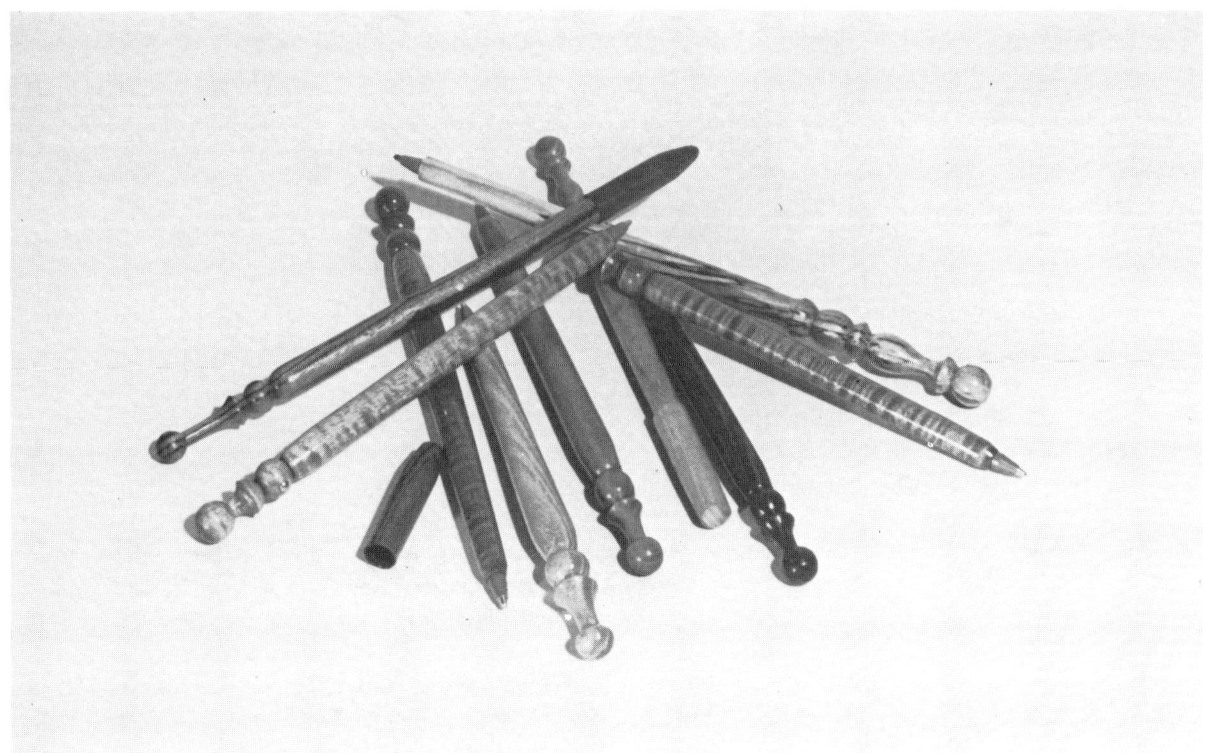

Illus.10. Matoesian pens

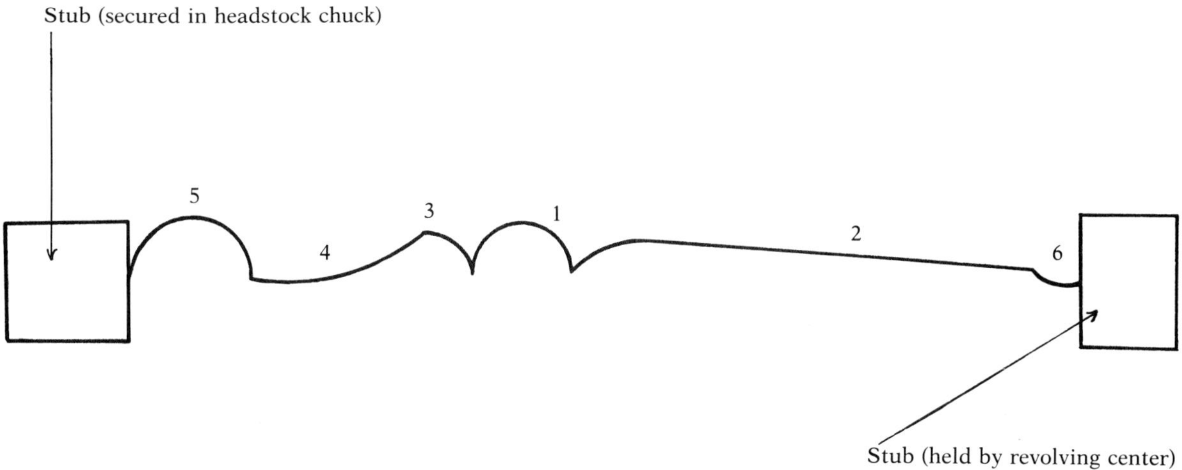

Illus.11. Schematic for turning pens

their caps from bocote, which is a Mexican hardwood. Spalted woods are also commonly used for pens. Since this project is so small, it can easily be made from scrap material.

For our purposes, we will focus on the basic design shown in Illus. 11. The numbers in the illustration refer to the different cuts Andy makes when he turns a pen. You will use this illustration as a frame of reference for many of the procedures in this project. Tasks 1 through 8 have to do with turning pens, and tasks 9 through 13 have to do with turning their caps. If you prefer to use desk-top bases instead of caps for some of your pens, turn to Chapter 3 for instructions.

One additional piece of information is important to note before starting the various project tasks. You can buy ball-point pen inserts from a number of mail-order suppliers. However, Andy prefers to use standard BIC pens since they are relatively inexpensive and readily available; to obtain the pen inserts, he simply pulls them from the plastic shafts that house them. Another reason he uses BIC inserts is that his friends and acquaintances can easily get insert replacements for their Matoesian pens should the ink run out.

TASKS:

1. While pens can be turned in a range of lengths and diameters, one of Andy's standard pen designs will be used for our purposes. Black walnut is generally the wood of choice for this design. It's also an excellent wood to use for learning how to turn a pen.

Using a band saw with a ½"-wide blade, cut 8"-long blanks that are ⅝" square. Since Andy enjoys the challenge of making pens in quantity, he generally will prepare anywhere from 25 to 50 blanks. I recommend that you also prepare a large number of them. You may break or ruin a few blanks while learning the various procedures—but more importantly, once you've learned how to make the pens, you're going to want to turn them in quantity.

On the headstock of the lathe, mount a ½" Jacob's chuck on a Morse taper. The chuck holds a homemade drive center that is ground from a ¼"-diameter straight router bit (Illus. 12). The two-pronged bit not only serves as the drive center, but also holds the blank firmly in place during turning. To make a similar two-pronged drive for your own use, you need a ¼"-diameter carbon-steel router bit. Hopefully, you will have one around your shop that's a bit dull and needs to be replaced. If not, a bit this size and type is relatively inexpensive.

Andy grinds the prongs to length and width on a bench grinder. You may want to refer to Illus. 12 to see how he shapes the bit on the grinder. You may prefer grinding the bit using a small Moto-Tool and grinding stone. *Whenever grinding, both Andy and I wear either safety glasses or a full-face shield. It's simply not worth the risk to do any grinding without eye protection.* After grinding the prongs to shape, you need to sharpen the ends for better wood penetration. You can easily do this on a whetstone.

In the tailstock, mount a revolving center. Even though you can use a stationary center and spread beeswax on the blank end, the ball-bearing type of revolving center is clearly preferable. This type of center is available in many varieties; some are inexpensive whereas others are very expensive. If you don't have a revolving center, it's worth your money to acquire one.

For proper alignment of the blanks between centers, it's a good practice to mark the midpoint on both ends of the blank. Make two straight lines from the corners that intersect. The midpoint is, of course, at the intersection. Andy is able to mount the blank, centered, without the benefit of markings, while the lathe is on. This is an old English turner's trick that is still used by production spindle turners. Although the ability to do this procedure comes with experience, it also has a lot to do with luck. Using the marked lines as guides, secure the blank onto the two-pronged bit first. Then bring the tailstock forward with the revolving center in place. Be certain the blank is secure before you begin turning.

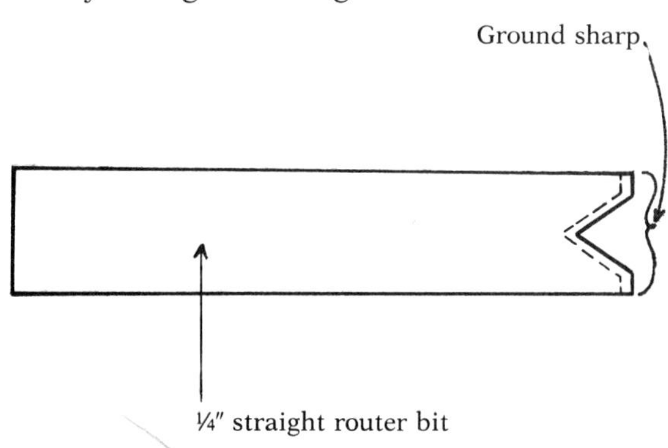

Illus.12. Drive center with shaped ¼″ router bit

For turning pens, Andy runs his lathe at 3500 rpm. After aligning a 12″ tool rest below the center line of the blank, he rough-turns the pen blank using a 1¼″ roughing-out gouge. The blank is rounded to a diameter slightly under ½″. Since the turned blank will be placed in the ½″ Jacob's chuck for turning, this diameter is critical. You should rough-turn all the blanks before moving on to the next task. Incidentally, if you're looking for a tool for roughing out spindles, you should consider the 1¼″ roughing-out gouge. It's a superb tool to use on a variety of turnings and, what's more, it's fun to use. As with any gouge, you should rub the bevel on the wood's surface (Illus. 13).

2. The next procedure involves drilling a hole in the blank that will eventually accommodate the ball-point pen insert. Andy places a three-jaw chuck on the headstock and puts the Jacob's chuck with its Morse taper into the tailstock. The three-jaw chuck holds the turned blank and the Jacob's chuck holds the bit for the drilling procedure (Illus. 14). To drill the hole, you need a standard-length 9/64″-diameter high-speed steel bit in addition to a longer aircraft bit with a 9/64″ diameter. To drill the hole, run the lathe at 2400 rpm.

This is the drilling procedure: First, secure the standard-length bit into the Jacob's chuck. With a turned blank secured in the three-jaw chuck, drill a shallow, true starting hole into the blank. You need to make this starting hole with the shorter bit because the longer aircraft bit is less reliable and tends to be easily deflected when contacting the wood's surface. You should make small starting holes in all the blanks. Even though this procedure seems to double the work, accuracy is critical. When you've made starting holes in all the blanks, replace the standard bit with the longer 9/64″-diameter aircraft bit. Repeat the drilling procedure, but this time allow the bit to penetrate its full length. Andy suggests that you retrieve the bit slowly so that it doesn't remove any additional wood from the hole.

By the way, the 9/64″-diameter hole will accommodate the pen insert's ink reservoir. The drilled hole should be deep enough to allow the entire reservoir to fit inside the blank. Later on, in the final procedure, you will drill another hole that is 5/32″ in diameter. This hole will hold the pen insert's front portion and point in place. More about this procedure later.

3. Now we will proceed with the decorative turning of the spindle. For this procedure, Andy again mounts the ½″ Jacob's chuck on the headstock and a ball-bearing center in the tailstock. You shouldn't use the two-pronged bit; place the rounded spindle (the end that's not drilled) directly into the Jacob's chuck instead. Then tighten the chuck around the spindle. The spin-

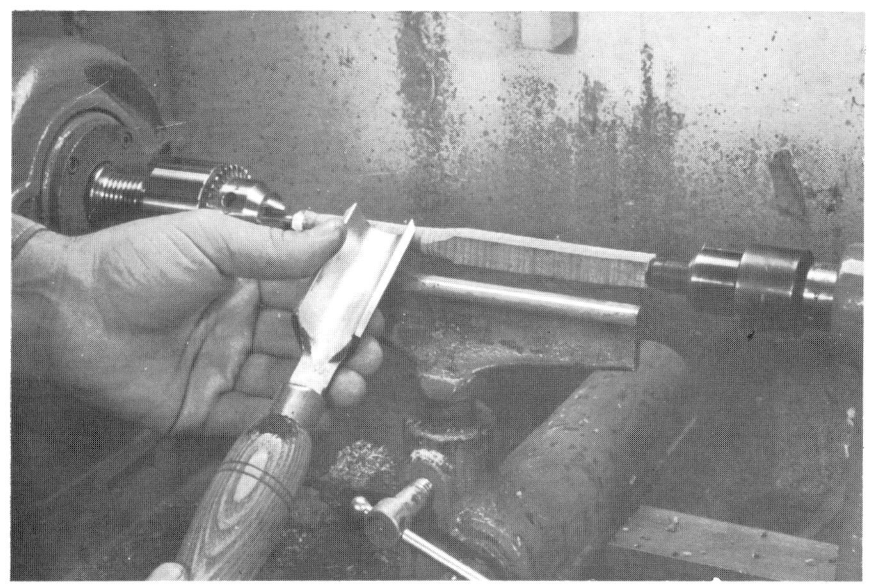

Illus.13. Andy rough-turning pen blank with 1½" roughing-out gouge

dle only needs to penetrate the chuck far enough to ensure a good grip by the jaws.

For a ball-bearing center in the tailstock, Andy prefers to use a cone-shaped or regular-cup center. He finds that the rotating spindle has more stability if the point in the cup is used for support. The point in the cup should extend at least ⅜" and penetrate the drilled-out portion of the spindle (Illus. 15). For decorative turning, Andy runs his lathe at either 2400 or 3500 rpm.

To understand or duplicate the various cuts made in the spindle, refer back to Illus. 11. This illustration is used as a reference for the tasks that follow. I would suggest you read through the material and orient yourself to both the tools and procedures prior to turning the spindle.

First, you cut the large bead (labelled 1 in Illus. 11) using a ⅜" gouge that is ground fingernail-style. Andy prefers the Henry Taylor Tools high-speed steel gouges because he thinks they're tougher than other gouges.

It's worth noting here how Andy sharpens his

Illus.14. Drilling insert hole in pen barrel

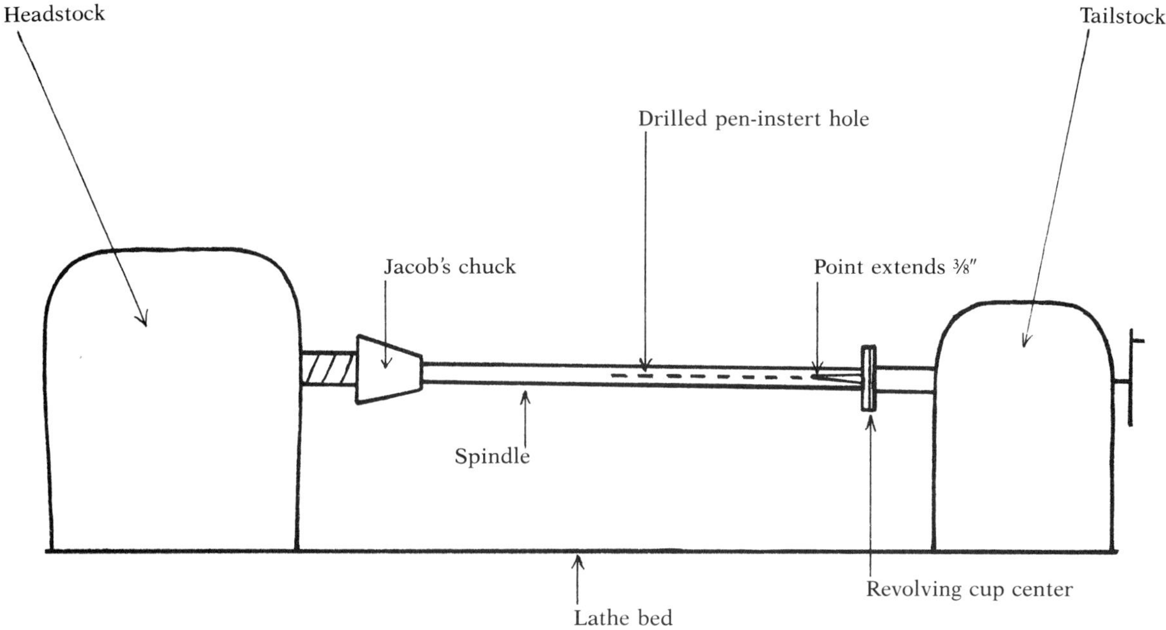

Illus.15. Spindle supported by cup center point

various gouges. He uses a standard electric bench grinder with a 1″ × 6″ friable pink wheel. The wheel is 60-grit aluminum oxide with a "J" bond. As mentioned earlier, this type of wheel, with its bond, is less likely to burn the tools. A sharpening jig is attached to his grinder that is based on a design developed by Russ Zimmerman (Illus. 16). Andy has modified the jig so that the large vertical block moves along a fixed support arm. A large hole is drilled into the block to hold the end of the turning-tool handle. The block is adjusted to allow the tool's bevel to rest on the stone, while the handle is held and supported by the vertical block. You can change the angle of the bevel by moving the block forward or backwards along the support arm. After the tool is properly aligned on the stone and in the block, it is secured to the arm by a large, threaded wood screw. With the grinder running, the gouge is lightly placed and rotated on the wheel. The fixed alignment of the tool assures a perfect, hollow ground bevel (Illus. 17). Although the jig procedures take some practice, this device is very effective for sharpening gouges. Best of all, it's homemade and inexpensive. I use it myself.

After the gouge has been properly ground, Andy hones and polishes the bevel on a 6″ buffing wheel. He cakes the wheel with Atlantic-brand greaseless compound in either a 60-, 80-, or 150-grit. When this is not available, he uses either an aluminum-oxide compound or black emery cake. Using a smaller buffing wheel, he also polishes inside the channel of the tang. You have to be careful that the sharp edge is not rolled or dulled during this procedure. Andy also rounds the bottom edge of the bevel where it joins the shank proper to make it easier to ride the bevel on the wood.

On the ⅜″ gouge that Andy uses for cutting beads, he has modified the channel of the shank. He has ground it wider in order to make a slightly shallower gouge. The shallow gouge enables him to turn beads more easily—so that he can get in tight on the bead, for example. While many turners use a skew to cut the beads, Andy finds the modified ⅜″ gouge easier to use. He fulcrums the tool off the rest when he makes his cuts. It's also critical to ride the bevel when making the cuts.

Back to cutting the bead (1 in Illus. 11). You should make the initial cut on the left half of the bead. Andy usually marks the dimensions of the bead with the long point of a skew. He scores the outer limits to make sure that the bead will be at least ⅜″ to ½″ wide. After making the initial cut to the left, make several additional cuts. Then make slices to the right to

Illus.16. Zimmerman's grinding jig.

balance the bead. Again, it's critical that you ride the bevel when making these cuts. This is not only done to avoid sanding, but also any scraping with the tool will cause the narrow spindle to vibrate and break; needless to say, it also dulls the tool.

4. The next procedure involves tapering the spindle. You should taper the barrel of the pen (2 in Illus. 11) using a roughing-out gouge. The roughing-out gouge must be lightly applied, and you should be certain to ride its bevel. While making this taper, Andy supports the spindle with his left hand. After he has made the taper, he planes the spindle with a ½″ skew chisel while supporting it with his left hand (Illus. 18). It's worth noting at this point that Andy seldom uses abrasive papers. The pens are ready for finishing as a result of the proper use of the tools.

5. You should cut the "V" notch (3) using the ⅜″

Illus.17. Andy grinding gouge on jig

Illus.18. Planing pen with ½" skew chisel and supporting with left hand

modified gouge. To make the notch, Andy uses the gouge almost as if it were a skew. After you have cut the notch, make the taper (4) using either a ⅜" or ¼" gouge. Andy seems to favor the ¼" Henry Taylor gouge, which tends to have more stability and less flexibility than most other small gouges.

After you've made the "V" notch (3) and the taper (4), cut the bead that terminates the taper (5). As you did with the first bead (1), use a ⅜" gouge. Because the bead is the top portion of the finished pen, you should carefully consider its diameter. Andy often makes it smaller than the first bead to give the overall design a more balanced appearance. When cutting this final bead, leave at least ⅛" of the spindle between it and the Jacob's chuck. The pen will be separated at this point; so some wood must be left in place.

6. Now move to the bottom end of the spindle. Again, using a ⅜" gouge, make a concave cut (6) that both tapers and reduces the diameter of the spindle. The taper is reduced at the tailstock end of the spindle. You need to be very careful with this procedure not to cut through the spindle wall and into the drilled insert hole. Andy tries to leave a wall thickness around the insert hole of at least 1/32". This diameter is, of course, at the outer edge of the taper.

7. After you've made the concave cut (6), separate the spindle at the headstock end. Using a ¼" gouge lying on its side with the bevel riding, separate the pen at the top of the bead (5) (Illus. 19). Place your left hand under the spindle while making this final cut. This way, you not only can provide support while you make the cut, but you can also catch the pen should it fall

Illus.19. Separating pen at top bead

off the cup center. There is usually a small residue of wood on the top of the bead after separation, but you can lightly sand it off as you prepare for finishing.

8. With the pen removed from the lathe, one final procedure remains. You need to drill a 5⁄32″-diameter hole into the barrel of the pen. This hole will accommodate the front portion of the pen insert and hold it in place. You should drill the hole before you separate the portion of the spindle that was supported by the cut center. Mount a 5⁄32″-diameter high-speed drill bit into a Jacob's chuck secured to the headstock or in a drill press. With the tool running, thread the spindle by hand onto the bit and drill out an area about ½″ long. Be certain that the pen is well aligned as you move it onto the drill bit. The excess wood that remains on the end of the spindle functions as a guide for this procedure.

After you have drilled the 5⁄8″ hole, you can separate the excess spindle from the pen using a band saw. Be careful when you are making the cut that the blade doesn't splinter the wood. As with the top portion of the pen, lightly clean up the cut edge with abrasive paper. Now slide a BIC pen insert into the barrel and check for fit. If it will not penetrate all the way into the barrel, cut a portion of the ink reservoir off with a pair of scissors. Replace the insert in the barrel and try out your new pen. Refer to Chapter 43 for finishing procedures.

9. Now for the caps. You'll be delighted to learn that it's much easier to turn them than it might appear. To look at them again, refer back to Illus. 10. Unless you are using turned bases for some of the pens you've turned, you need to make a corresponding number of caps. (As previously mentioned, instructions for turning bases are presented in Chapter 3.) Andy prepares blanks for the caps that are approximately 5⁄8″ square and 7″ long. He uses the same wood for the caps as he used for the pens. By the way, you can get two caps from a 7″-long blank.

Mount the blanks between centers with a Jacob's chuck and the two-pronged router bit used as a drive. Place a revolving center in the tailstock. With the lathe running at 3500 rpm, Andy rough-turns the blanks using a 1¼″ roughing-out gouge. Now turn the spindles down to a diameter of approximately 7⁄16″. After you've rough-turned the spindles, cut them into two equal sections. A band saw is the tool of choice for this procedure. By the way, hold the round spindles firmly while cutting them on the band saw. The blades of the band saw tend to grab round objects and tear them from your hands. You may want to consider making a small jig for this sawing procedure.

10. Remove the two-pronged drive bit from the Jacob's chuck and secure a spindle section in its jaws. Since you're using a ½″ Jacob's chuck, it's critical for you to have turned the spindles down to a diameter of at least 7⁄16″. If the spindles are not less than ½″ in diameter, they won't fit properly in the chuck. You should insert the spindle section into the chuck just far enough for the jaws to grab and hold it firmly in place. When the lathe is turned on, you will find that the spindle section will probably be running a bit off-center. When this happens, Andy uses a 1¼″ roughing-out gouge, on its side, to true up the spindle.

11. After you have trued up the spindle, drill it out so that it will fit over the pen barrel. The final diameter of the pen dictates the size of the hole that you need to drill into the cap. According to Andy, high-speed steel bits in diameters of ¼″, 5⁄16″, or 3⁄8″ are adequate for most pens. If none of these sizes works, check the pen diameter with a vernier caliper.

To hold the drill bit, place another Jacob's chuck on a Morse taper in the tailstock. Since Andy doesn't like to change speeds on his lathe, he leaves it at 3500 rpm for the drilling procedure.

Depending upon how you use your lathe, you can drill the caps with the tailstock stationary or the bit moved into the wood using the drive wheel. Another option is to slide the entire tailstock forward and backwards to drill the hole. Andy drills the hole to an approximate depth of 1½″. After you've drilled the hole, insert a pen and check for fit. If it's a good fit, you may not need to do the next procedure (Illus. 20).

12. For some pens, you will need to taper the drilled hole inside the cap. Since the pen barrel usually has a taper to it, sometimes you must

Illus.20. Drilling hole in pen cap

also taper the cap hole for a good fit. Andy uses a tapered reamer for this procedure. Tapered reamers are used by machinists and come in a variety of types and sizes. For this procedure, a #5 tapered reamer of high-speed steel is necessary. This size is adequate for most caps. Since the taper on the reamer is usually very slight, Andy cuts off the first inch or so of the reamer (Illus. 21). The grooves machined into the tapered reamers are very sharp, and produce a fairly smooth surface inside the cap.

To use the reamer, you must first secure it into the Jacob's chuck on the tailstock. If you desire, you can move the reamer into the cap and rotate the lathe by hand. The reamer is sharp enough so that it will cut as the cap revolves. The other option is to turn the lathe on and move the reamer into the cap. You may feel more confident if you make the taper while you turn the lathe by hand.

13. After he has made the taper in the cap, Andy uses a ½″ skew chisel to plane and thin down the outside wall of the cap. After this procedure, the wall of the cap should be between ⅛″ and ³⁄₁₆″ thick. On some woods, especially the exotics, you can thin the wall to almost-paper thickness.

14. Now remove the cap from that portion held in the Jacob's chuck. Using a ⅜″ fingernail gouge, cut a slight convex taper as you separate the cap. Lightly sand the end of the cap to clean it. Then refer to Chapter 43 for finishing products and procedures.

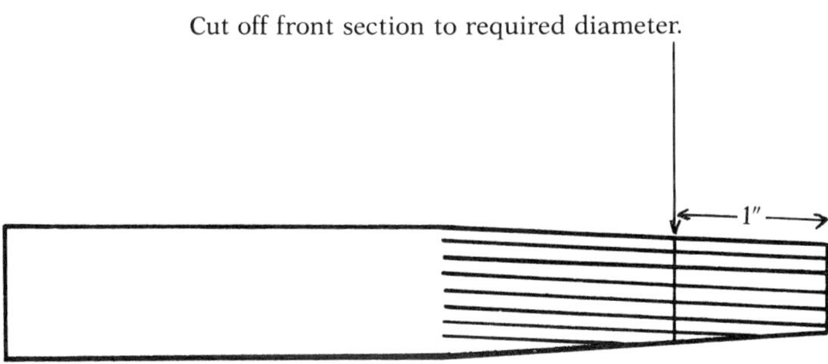

Illus.21. Pen-cap reamer

3
BALL-POINT PEN BASES

This is the perfect complementary project for the ball-point pens presented in Chapter 2. As with the pens, the design for these formal desk-top bases is one frequently turned by Andy Matoesian. A number of projects in later chapters present some options for bases that are meant for commercial pens. You may want to look over some of these designs and integrate certain elements into this project. One of the enjoyable parts of turning Andy's pen bases is

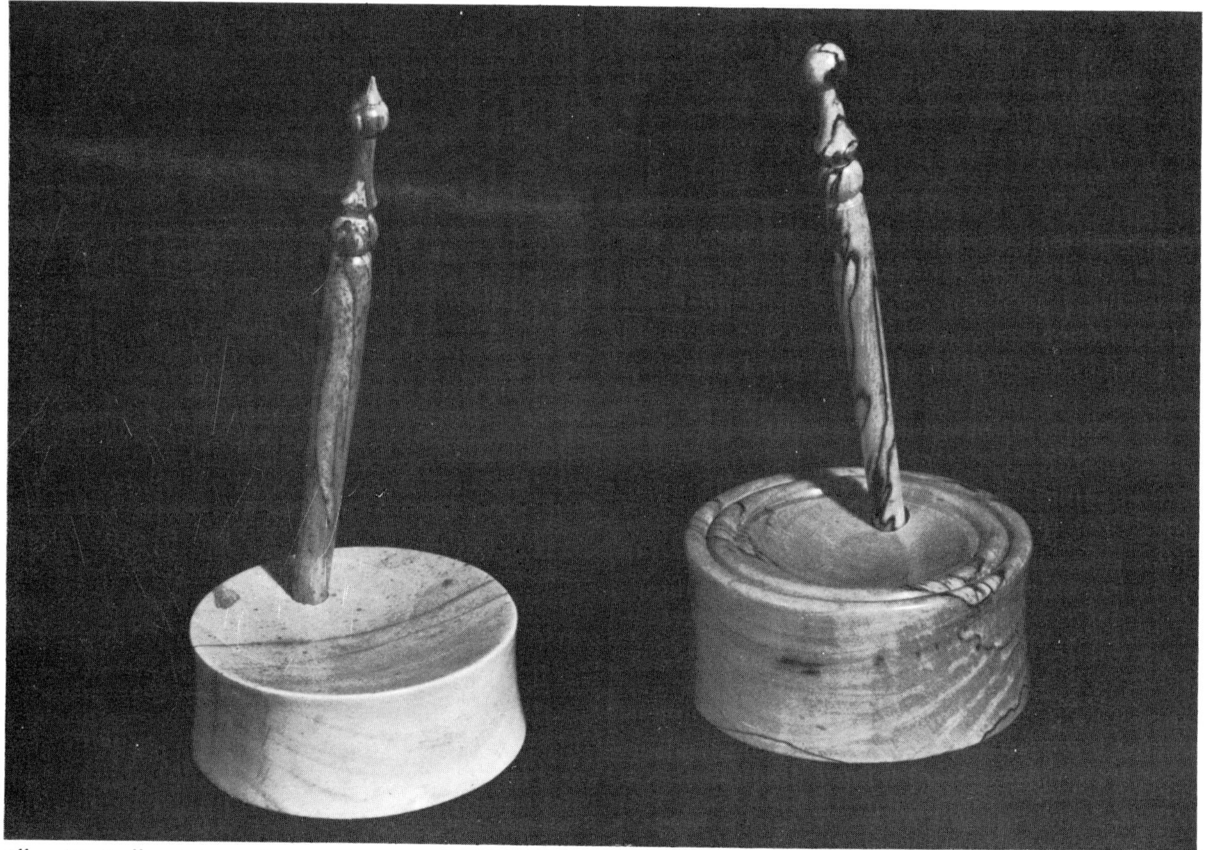

Illus.22. Ball-point pen bases with pens

that you can use gouges if you so desire. The project also presents an interesting homemade scraper that is used for making beads. Purist cutter that he is, Andy refers to this homemade device as a "cheater."

You can turn this project using either double-face tape as described in Chapter 1 or a screw center. And you can use either of the homemade chucks described in Chapter 1 or a commercial chuck. For purposes of demonstrating how to do the project, Andy uses a homemade steel chuck that he designed himself.

It's worth noting that it would benefit the hobbyist woodturner to find a local machinist and get to know him well. In addition to the screw center that is used with this project, Andy and I have our faceplates and other accessories made by a friend who is a hobbyist machinist and mechanic. Not only is this an excellent way to save money, but you can have items made the way you want them. For example, both of our lathes have similar spindle diameters and thread sizes. We locate large 1 × 8 threaded nuts and 5⁄16″-thick steel and have our friend make our faceplates for us. He welds the nuts to the steel and trues the plate on a metal lathe. The only remaining task for us is to drill screw holes and then countersink them on the faceplate surface. We countersink the holes to allow a space for the wood fibres to gather when the screw penetrates the wood. By doing this, we can be sure that the faceplate and the block's surface will be flush when the screws are in place.

In addition, when you need a special tool, a local metalsmith or machinist can be invaluable. For example, following Andy's design, a local metalsmith made the "cheater" that Andy uses for this project. You would do well to find a metalsmith or machinist in your area so that you can take advantage of his expertise and skills.

TASKS:

1. The initial task is to pattern and cut blocks for the pen bases. You need to use stock for the blocks that is at least 1¼″ (5/4's) thick, and it should be similar to the wood you used for the pens. Pattern the blocks to a diameter of at least 3¼″. As mentioned in Chapter 1, a plastic template with a range of precut circles is an effective device to use for patterning blocks. The tool of choice for cutting the blocks is a band saw with a ¼″-wide blade.

2. If you plan to use a screw chuck, you must drill an undersized hole into the block for mounting on the chuck. The diameter of this hole is determined by the diameter of the chuck screw. You want the chuck screw to firmly hold the block while turning; therefore, the hole must be appropriately undersized. Also, the drilled hole in the block should not penetrate the top surface. For example, if you decide to have the top surface of the base slightly concave, leave at least ⅜″ of wood between the end of the screw hole and the top surface of the block. Incidentally, while there are devices—such as a plastic circle template—to help you locate the exact middle of a round block, a visual approximation is acceptable for the project.

When you mount the block on the screw of the chuck, the screw may be too long for the depth of the drilled hole, depending on the type of chuck being used. Should this occur, make a small shim, drill a hole in it, and thread it on the screw. With an appropriately thick shim between the face of the chuck and the block, the length of the screw should be sufficiently reduced. Then you should be able to thread the block tight against the shim's surface for turning (Illus. 23).

If you're uncertain about turning a block while it's mounted on a screw chuck, place a revolving center in the tailstock and bring the assembly forward. You can either force the point of the revolving center into the block's surface or place a piece of dowel between the two. The dowel will prevent any marking of the block's face surface. You may want to refer back to Illus. 9 to review this setup.

3. For initial shaping of the base block, Andy uses a 1¼″ roughing-out gouge. After the block is turned true, he uses a ⅜″ gouge to concave the side of the block. You can also accomplish these tasks using a 1″ roundnose scraper.

4. After realigning the tool rest to the front of the block, Andy cuts one or two beads into the front surface using his "cheater" (Illus. 24). The bead cutter is made from a ⅜″ × ⅜″ machine-

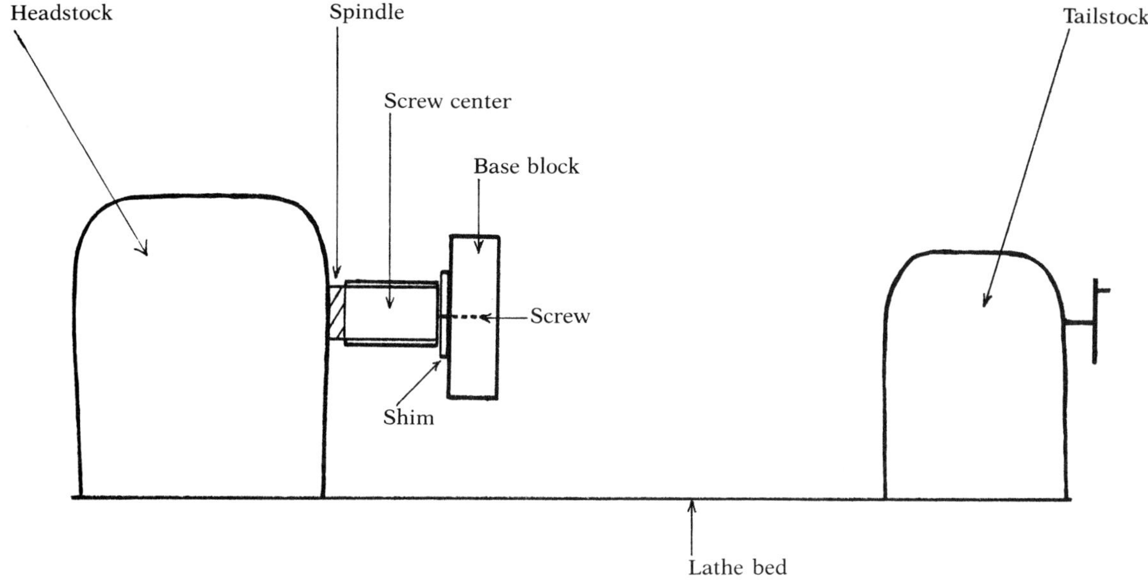

Illus.23. Screw chuck, shim, and block

tool bit of high-speed steel. The bit is brazed to some mild steel that serves as the shank. Then the bit is ground on the round edge of a grinding wheel to give it the shape that is needed. The scraping edge of the bit is then ground, leaving a good wire edge (Illus. 25).

To effectively use the bead cutter, the lathe needs to be running at a high speed. Also, as the bit is placed against the wood, you need to wiggle it from side to side while making the bead. You can make one or more beads on the top surface of the base, depending upon your own taste or preference.

After forming the beads on the top surface, Andy squares the surface between the beads using a square-nose scraper. If he wants a concave effect, he uses a roundnose scraper. At some point, all good cutters must succumb to using those dastardly scrapers.

5. Now it's time to drill a hole in the top for the pen and to fill in the screw-chuck hole in the bottom of the base. You should drill the hole at a slight angle, depending upon the finished diameter of the pen. A ⅜"-diameter brad-point bit is adequate for most pens. The slight angle of

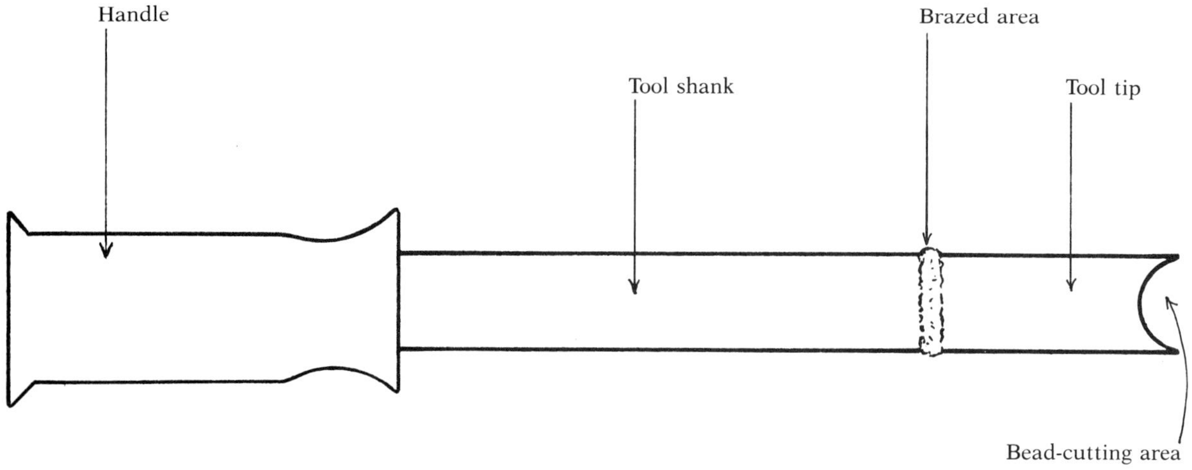

Illus.24. Andy's cheater for cutting beads

Illus.25. Cutting beads with cheater

the drilled hole is necessary so that the pen will rest in the base at an angle (Illus. 26). While using the 3/8″ bit, drill another hole over the screw-chuck hole in the bottom of the base. Using a 3/8″-diameter plug cutter or a piece of 3/8″-diameter dowel, make a plug and glue it in place. However, you may prefer to use wood filler instead. Using a range of abrasive papers, sand the bottom of the base. Then refer to Chapter 43 for finishing products and procedures.

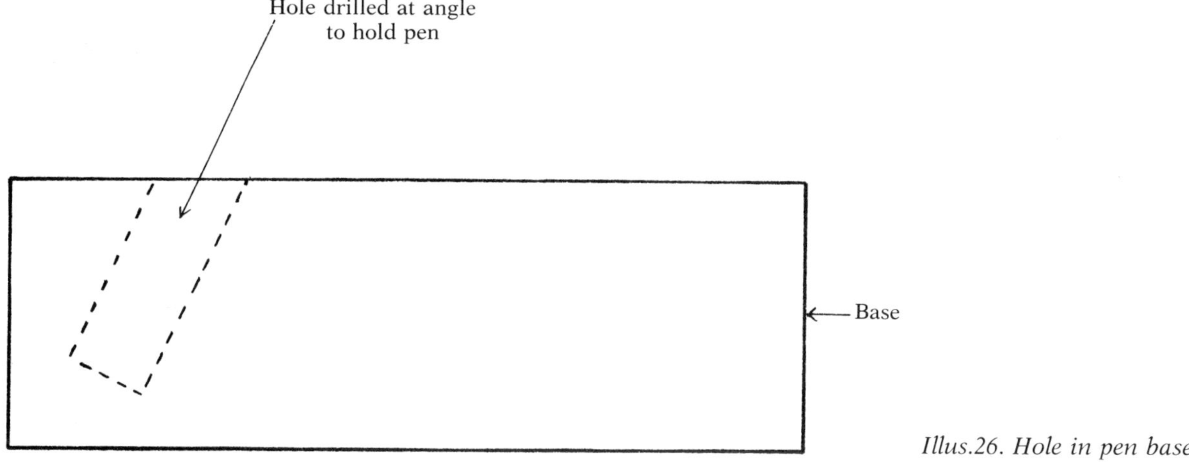

Illus.26. Hole in pen base

4
SALT & PEPPER SHAKERS

While salt and pepper shakers are a rather common turning project, these shakers are a bit different. For one thing, they are true shakers—not grinders that require internal mechanisms. Additionally, you can turn them to a variety of shapes using a rather simple homemade chuck. You will need to acquire stoppers for holding the salt and pepper inside the shakers; however, stoppers are readily available in plastic, rubber, or cork, and in a range of diameters. I generally use 1″-diameter plastic stoppers.

To differentiate the salt shaker from the pepper shaker, I usually will vary their tops. On occasion, I will vary the height or the diameter of the two shakers. Since I make only one hole in each top to permit the salt and pepper to shake out, I generally will vary the diameters of the holes. The salt shaker always has the larger hole. However, you may prefer using other design options that will help the user discriminate one shaker from the other. These design issues make salt and pepper shakers an interesting and rather challenging project. You can take a somewhat mundane pair of functional kitchen items and explore a whole range of design possibilities.

Salt and pepper shakers can be made from almost any wood, but, for obvious reasons, it's best to avoid the potentially toxic imported woods. The salt and pepper shakers in Illus. 27 have been turned from Osage orange, or hedge. Actually the hedge had been used as a fence post for many years—thus the brilliant golden color of the wood. I use black walnut to demonstrate this project. You can, however, use oak, cherry, or fir—or any other wood. If you have a preference for matching wood, you may want to turn a set of shakers from the same wood that was used for your dining-room table. How about maple and maple? In addition to the solids, you can also laminate contrasting woods and even use strips of veneer. There are many design options you may want to consider as you

Illus.27. Salt and pepper shakers

prepare your turning blocks. More about this shortly.

With this project, not only can you try different designs, but you can also experiment with a range of tools. You can use gouges or scrapers—the choice is up to you. This may be the project where you develop some new techniques and skills using a new method or an unfamiliar tool.

TASKS:

1. I often make the shakers at least 4″ high and about 2½″ in diameter; so I need to use stock that is at least 2¾″ (11/4's) thick. While you may not have access to black walnut in this thickness, frequently you can obtain 4″ × 4″ posts in oak or fir. I've turned shakers from basswood because it's usually available in very thick stock. If you purchase the lower grades, they are often laced with mineral staining and other natural aberrations. The required dimensions for the project also make hedge fence posts a good alternative. I'm confident that, with a bit of ingenuity, you will discover other sources of material that will meet your dimensional needs. If all else fails, find some dry fireplace logs and use them.

You need to cut the block with the grain. This not only gives the finished piece an unusual appearance, but with the block mounted on a chuck and the tailstock brought forward, it also gives you the opportunity to do some spindle turning. If you plan to use solid blocks for the project, pattern and cut them to a 4″ length and a 2¾″ diameter. As mentioned earlier, you may want to vary the height or the diameter of the two shakers to help the user discriminate one from the other. You need to address these kinds of issues while you are preparing the turning blocks.

It's best to cut the blocks on the band saw; but be careful because, in addition to the risks involved in cutting with the grain, blocks this high and narrow can also present some problems. *Use a ¼″-wide blade and, by all means, take your time and keep your fingers away from the blade.*

If you would rather not use solid blocks for the project, there are any number of alternatives. One option is to glue together three pieces of 1″ (4/4's)-thick stock that is 4″ in length or longer. If you want to significantly change the overall design, laminate four pieces of stock for the block. This will enable you to turn a larger diameter at the midsection of the shaker. You can also use some combination of contrasting woods—such as walnut, maple, cherry, and oak—when you glue a block together. With this procedure, you can use scraps that you may have around your shop. Another option is to glue strips of contrasting veneer between the various pieces. You can use one veneer strip or a number of veneer strips to create some interesting and unusual effects. No matter which option you choose, be sure to use enough wood glue, clamp the assembly securely, and allow the glue to properly dry. Then proceed with patterning and cutting the blocks for turning.

2. The next task in the project involves drilling 1″-diameter holes into the two blocks. The holes should be drilled almost through the blocks. I leave about ¼″ of wood between the hole and the block's top surface (Illus. 28). This hole is used to hold the turning block on a homemade chuck during the turning process. Of course, it's final use is to hold the salt or pepper. The diameter of the hole determines the size of the stopper. If you prefer storage areas in your shakers that are larger than 1″ in diameter, use a larger-

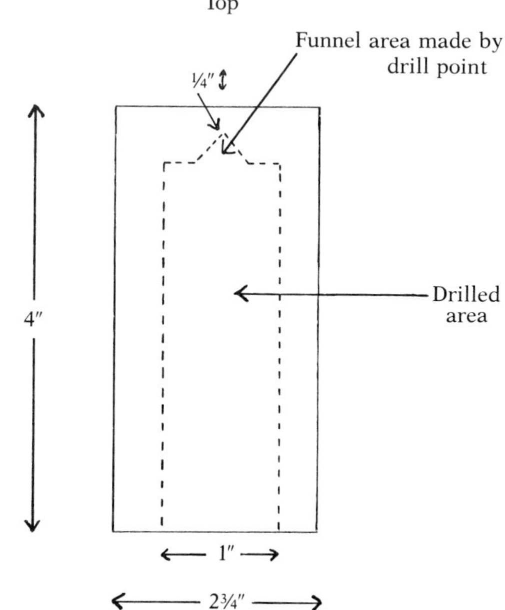

Illus.28. Shaker block

diameter bit. But check to be certain that stoppers are available in the diameter you select.

To drill the holes, I use a 1″ multispur bit secured in the chuck of a drill press. You need to mark the middle of the block's end surface, where the point of the bit will enter. Because your blocks may not be perfectly round, you may have to "eyeball" or approximate the midpoint. You may want to clamp the blocks while you drill the holes. Be certain that the point of the multispur bit doesn't penetrate the top surface. As I just mentioned, I tend to leave at least ¼″ of wood between the drilled hole and the top surface of the block.

I exploit the small, tapered holes made by the point of the multispur bit. You can use them as funnels for the salt and pepper when you shake them from their vessels. Later you will simply drill a final hole through the top surface of each block that will penetrate the tapered hole.

3. As I've indicated, I prepare a homemade chuck that is mounted on a 3″ faceplate to turn this project. Illus. 29 shows a dimensional drawing of the chuck. To make the chuck, I use a 4″ × 4″ square from either an oak or a fir post. You may prefer another wood in the required dimensions. Using a school compass, pattern the block to the largest possible diameter. I like the finished chuck to have a diameter of at least 3½″ where it attaches to the faceplate. You should pattern and cut the block across the grain, if possible, so that the faceplate screws will have more holding power when you secure them to the block. Unless you use exceptionally long screws, you won't be able to get much holding power with end grain.

After cutting the block, attach it to a 3″ faceplate. You may want to refer to Chapter 1 for a brief discussion of the square-recess screws that I use; they're ideal for faceplate work. Mount the assembly on the lathe and true it up using the tool of your choice. Or you can secure a revolving center in the tailstock and force it against the block; this will give you both support and stability while turning.

After you have turned the block round, begin cutting it to a long taper. Refer to Illus. 29 again for some suggested diameters along the length of the taper. As you might guess, the shaker blocks, by means of their 1″ holes, are forced onto the taper for turning. Thus, for good holding power, you don't want the tapering to be too radical. It needs to penetrate well inside the block hole for proper support and holding power. What primarily holds the block on the chuck taper is friction and support from the tailstock. A vernier caliper is an effective device for checking the various diameters along the taper. Before removing too much wood from the taper, periodically check the fit of a block hole on the taper. When you have a good fit and sufficient penetration of the taper in the hole, stop turning.

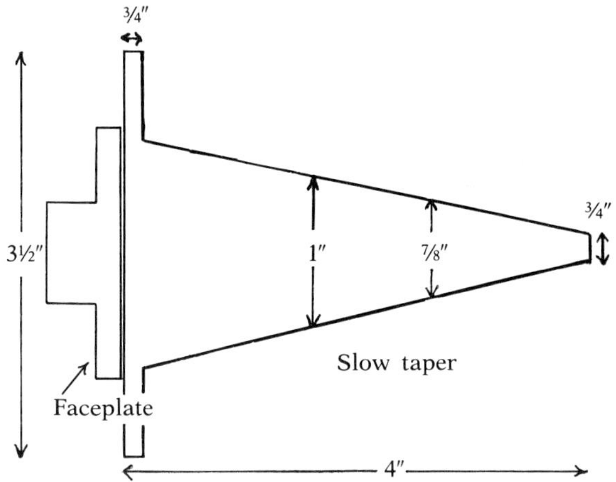

Illus.29. Homemade chuck for turning shakers

4. Mount one of the blocks on the tapered chuck, and move the tailstock with a revolving center into the block. It's too risky to turn the blocks without the tailstock in place. You want the revolving center to penetrate as near the middle of the block's surface as possible. The indentation that it makes in the top surface is where you will eventually drill a hole for the salt or pepper to be removed from the shaker. If everything works according to plan, the hole should penetrate the funnel made by the multispur bit.

For turning the blocks true, you can use either a 1¼″ roughing-out gouge or a 1″ roundnose scraper. Before you do very much shaping, you may want to refer back to Illus. 27 to see the finished project. You may want to significantly alter the design or add beads and coves to it. I happen to like the simple lines of this design, but your taste may be different. Use a minimum of tool pressure when you are rough-turning or shaping the block. Too much pressure can cause the block to slip on the chuck taper.

To distinguish between the salt and pepper shakers, I usually vary both the diameter and the width of the tenons on the top of the shakers. On the salt shaker, I tend to make the tenon ¼″ long and ⅞″ in diameter; on the pepper shaker, I will then make the tenon 3⁄16″ long and about ½″ in diameter. However, these dimensions are arbitrary; so you can modify them as desired. The main point is that these are two easy ways to turn the blocks so that the user can quickly distinguish between the two shakers.

Using your tool of choice, begin rolling the front portion of the block towards the revolving center. I generally use either a ½″ roundnose scraper or a ⅜″ fingernail gouge for this procedure (Illus. 30). After doing some initial shaping, make the cut for the tenon using a ½″ square-nose scraper. I make the cut a bit wider than necessary because the front surface of the tenon will need to be cleaned up, which always reduces the length of the tenon. To check the accuracy of the tenon's diameter, use a vernier caliper. Or you may prefer to simply "eyeball" the diameter until it looks proportional to the base of the shaker. The main thing with this design is that you vary the tenons on the two shakers. Continue shaping the front portion of the shaker and cleaning up the edge and shape of the tenon until you're satisfied.

Now you need to taper the bottom surface of the shaker so that the stopper does not extend beyond the bottom when in place. The type of stopper you will use determines the degree of taper (Illus. 31). For example, a cork stopper tends to extend quite far, even with a portion of it penetrating into the reservoir hole. A plastic stopper, on the other hand, extends only the width of the lip that surrounds the base of the stopper. This is one of the reasons I prefer using the plastic stoppers.

You can make the tapered bottom using the point of a skew or a parting tool. Beginning near the edge of the shaker base, move the point of the tool inward and slightly upwards to cut the taper. Be careful that you don't cut into the chuck. You can leave some wood around the chuck as you make the cut. You can easily break off this excess wood and sand the area when you remove the shaker from the chuck. Be sure the taper is sufficient to accommodate any extension of the stopper.

5. You may want to refer to Chapter 1 for a brief discussion of some sanding procedures that I find useful. However, depending on the quality of your tools and how effectively you've used them, you may or may not need to use abrasive paper. If sanding is required, do it before the tailstock is removed. You will want to maximize the support it provides for the sanding process. If the wood grain runs parallel with the lathe bed, turn off the lathe and sand the surface while rotating it by hand. Prepare the surface for finishing.

6. Remove the tailstock and realign the tool rest in front of the shaker. Using a very light touch, clean up the top surface of the tenon. Using too much pressure with the tool may move the block off-center on the tapered chuck. I usually

Illus.30. Shaker on homemade chuck

do this task with a ½″ square-nose scraper. You can sand the top surface of the tenon after you have removed the shaker from the chuck.

7. After you have turned and sanded both shakers, you need to drill holes through the middle of the tenons that will allow the salt and pepper to be removed from the reservoir. For the salt shaker, I drill a ⅛″-diameter hole using a high-speed bit. You should drill the hole at the indentation that was made by the point of the revolving center. If everything has been properly centered, the drill bit should penetrate the internal tapered hole left by the multispur bit. This tapered hole, in turn, will act as a small funnel to facilitate the flow of the salt from the shaker. You should drill a smaller hole through the tenon of the pepper shaker. I use a 1⁄16″-diameter bit, but you may prefer to use a larger one. If possible, it's best to drill the shakers with a drill press.

Using a range of abrasive papers, sand the bottom tapered area and the area around the stopper hole. Place the stoppers in the holes and test for fit. Refer to Chapter 43 for products and procedures to use in the finishing process. Since you won't be finishing the insides of the shakers, it's not necessary to use a nontoxic finish.

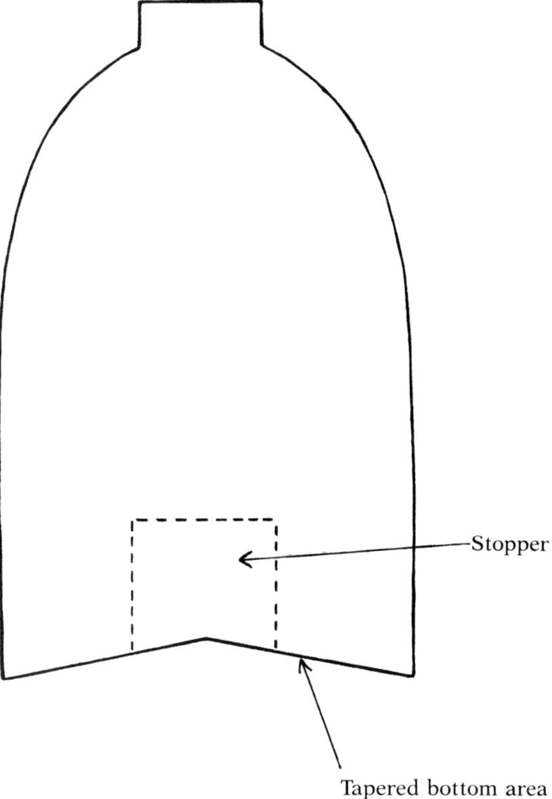

Illus.31. Tapered shaker bottom

5
TEXTURED BOWL

Upon examining some 3½″ (14/4's)-thick black walnut stock, I discovered a network of checks throughout the wood. The wood has been air-dried for a number of years at a local sawmill and then lay in my shop for a while longer. The checks were clearly from improper drying procedures that left the stock almost worthless. On occasion, when there is but one serious check in a block, I will cut it in two pieces along the check, glue a piece of veneer between the pieces, clamp it together, and eventually turn the block. But when there is an entire network of checks, it's another matter.

Since I'm the kind of person who does not easily throw away wood, especially 3½″ (14/4's)-thick black walnut, I reflected on possible uses for the checked piece, and the idea of trying an unusual texture on a turning began to peak my interest. Since potters create absolutely beautiful textures on some of their pieces, I decided to add some further texturing to the checked walnut block. After experimenting with various methods and textures, I came up with the process that's used in this project.

I turned the bowl in Illus. 32 from a badly checked piece of 3½″ (14/4's)-thick black walnut, but I frequently turn and texture bowls of varying sizes from sound stock.

TASKS:

1. To turn this project, I would select a chunk of wood that is at least 2″ (8/4's) thick and not particularly attractive. Better yet, you may have a block that is seriously checked, but examine the wood carefully to be sure it won't fly apart when you're turning it. Since you will be turning the block on a 3″ faceplate, you need to use something that will give you a diameter of at least 4 to 5 inches. You may want to make your bowl the same shape as the bowl in Illus. 32. Now pattern and cut the block.

2. Although a number of methods for mounting blocks for faceplate turning are discussed in

Illus.32. Textured bowl

Chapter 1, I would recommend that you secure the faceplate directly to the block for this project. You may want to try the square-recess screws that were discussed in Chapter 1; on larger blocks, I always attach the faceplate directly to them. To add support and security, I also place a revolving center in the tailstock and bring it forward into the block. I leave it in place even while I'm removing wood from inside the block. A large block flying off a faceplate at 2500 rpm can be extremely hazardous to you or anyone else who happens to be in the shop. After securing the faceplate to the block, place the assembly on the lathe and bring the tailstock and revolving center forward.

3. With the tool rest properly aligned, turn the block round. You can true it using either a ⅜"-deep bowl gouge or a 1" roundnose scraper. When you are making a textured bowl, you obviously don't have to worry about a fine, smooth surface. You don't even have to be concerned about cross grain or mouse. As a matter of fact, you don't even need to use abrasive paper until the texturing has been done. More about this shortly.

After you've turned the block round, shape it as desired. You may want to duplicate the shape of the bowl in Illus. 32, or come up with a shape of your own. Do the shaping with a tool of your choice. I generally continue using the 1" roundnose scraper. As I've indicated, I usually turn at 2500 rpm. You may prefer a bit more speed or feel more secure with a slower speed. Given the nature of this turning project, you may want to try out various tools and lathe speeds; this is one of those projects where you can experiment with your turning techniques without worrying about damaging the block.

4. After you have shaped the bowl, realign the tool rest so that you can remove the wood from the inside of the block. I use a 4" tool rest for this procedure because it allows me to keep the tailstock and revolving center in place. For quick removal of wood, use a freshly ground ½" or ¾" square-nose scraper. Unless I'm trying to show off, I seldom turn bowls with paper-thin walls. With a textured bowl, you will definitely want to leave a wall that is at least ⅜" thick. If you have some very serious checks in the surface, you may want the wall to be even thicker. I also tend to make these bowls rather shallow. I enjoy the weight of the wood in a turned piece. With a 2" (8/4's)-thick block, you may want to remove only about 1" of wood from inside the block.

5. When you have removed the bulk of the wood from the inside of the block with the square-nose scraper, cut through the pillar that has developed under the revolving center by using the corner of the scraper. Be careful with this procedure because the separated pillar can fly out of the rotating bowl and strike you in the face. *This is a time when you should be wearing protective eye or face shields*. When you have removed the pillar, slide the tailstock out of the way and replace the 4" tool rest with a larger one. With the longer tool rest, it will be easier to shape the inside of the bowl.

Using a 1" roundnose scraper, shape the inside wall to conform to the outside wall. I also round the inside wall so that it flows into the bottom surface. With a sharp scraper, you can also clean up the bottom surface. As you shape the inside, monitor the wall thickness and try to maintain a thickness of at least ⅜". A good device for checking wall thickness is a double-ended caliper (Illus. 33), which you can make

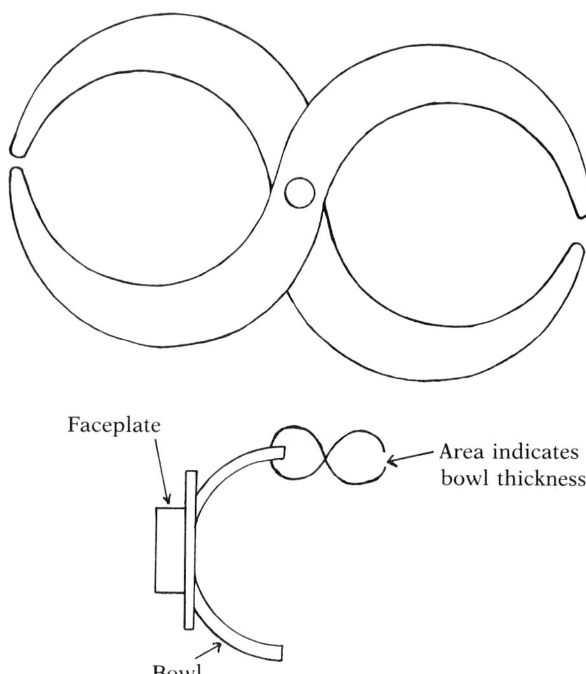

Illus.33. Double-ended caliper

yourself or buy from a turning-supply house. You can also check the thickness of the wall by using your thumb and index finger to feel along the wall. Unless the project requires precision, this procedure is usually adequate.

When you have turned the inside of the bowl to your satisfaction, square the top edge of the bowl using a freshly ground square-nose scraper. I make the edge at least ¼″ wide. Since I don't texture this edge, it needs to be cut clean and smooth. Of course, I don't texture the inside of the bowl either. You will want to sand the inside surface after you've completed the texturing procedure.

6. To texture the surface of the bowl, I use a rather unlikely tool—a 14″-long half-round wood rasp with a 3″ curved, bent, and tapered tang (Illus. 34). Curiously enough, I use the end

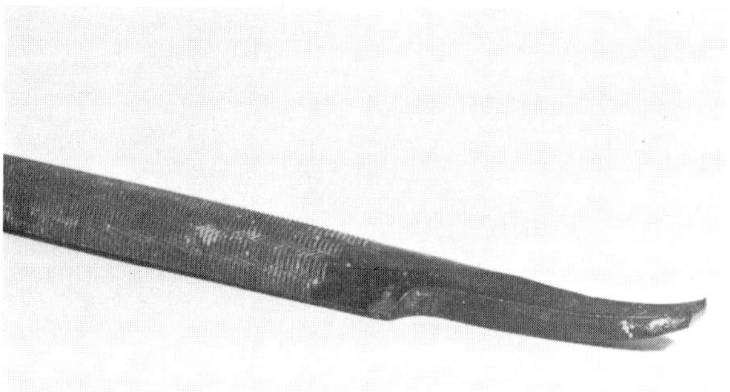

Illus.34. Fourteen-inch rasp for texturing

of the tang to texture the bowl's surface. I grind the end of the tang square, leaving a sharp edge on the bottom of the bend. The tang end should be sharp so that it will cut easily into the surface of the wood.

Realign the 12″ tool rest next to the bowl and slightly above the center line. The rasp is pointed downwards during use, similar to the way you would use a scraper. Lay the rasp on its side and force the sharp end of the tang into the wood (Illus. 35). I run the lathe at 2500 rpm for this procedure. You want the tang to cut a groove that is at least ⅛″ deep and circles the bowl. After you cut the first groove, make a second cut that is approximately ¹⁄₁₆″ away from it. Repeat this process over the entire surface of the bowl. You will get some wood strands at the cross-grain areas of the block, but you can easily cut them off with abrasive paper.

To prevent the grooves from becoming rough, lightly regrind the tang of the rasp periodically. As I've said, do not texture the flat top edge or the inside of the bowl. The cleanly cut top edge is an attractive contrast to the textured surface and it enhances the overall appearance of the project. After you've made all the cuts, shut off the lathe and examine the bowl's surface. You may find a few areas where the space between the grooves is too great. Also, you may find grooves that need to be cut deeper. Turn on the lathe and redo these areas.

After I have textured the surface, I use abrasive papers just to cut off any wood strands that may be sticking up from the surface. I generally use 100- and 150-grit abrasive paper and then, as a final procedure, go over the entire surface with 220-grit. You should also lightly sand the top edge and the inside of the bowl.

7. When you have completed the sanding, remove the faceplate from the block. Although you can fill the faceplate-screw holes with wood filler, I prefer using homemade wood plugs. Using a ⅜″ diameter plug cutter mounted in the drill press, cut plugs from the same wood you used for the bowl or from a contrasting wood. After you have made the plugs with the cutter, you can saw them from the board on the band saw, using the fence. Holes are drilled into the faceplate-screw holes with a ⅜″-diameter brad-point bit. After placing a bit of wood glue in the holes, tap the plugs in place. Allow the glue to dry. If necessary, use a large belt sander to sand the plugs flush with the bottom surface. Finish sanding the surface with a series of finer abrasive grits. Now refer to Chapter 43 for finishing procedures.

Illus.35. Rasp tang texturing bowl

6
CANDLEHOLDERS

Candleholders are always an enjoyable and worthwhile project to turn on the lathe. Although more elaborate candleholders are presented in Chapter 39, this simple design requires a minimum of wood and takes a minimum of time. There is always a need for more candleholders in a household, especially ones that are practical and attractive yet also inconspicuous. You can use these simple candleholders in a formal dining setting without overwhelming everything else on the table.

With this project, you get the chance to explore various procedures for securing blocks to faceplates or chucks. For example, the candleholders are ideally suited for using double-face tape or one of the homemade screw chucks described in Chapter 1. If you use double-face tape, you can also get some experience drilling holes using the tailstock and a Jacob's chuck.

This is a project that you can make in quantity, while using only a small amount of wood. The candleholders are 1″ (4/4's) thick and approximately 3″ in diameter.

TASKS:

1. As indicated, you will need stock that is at least 1″ (4/4's) thick for this project. Given the

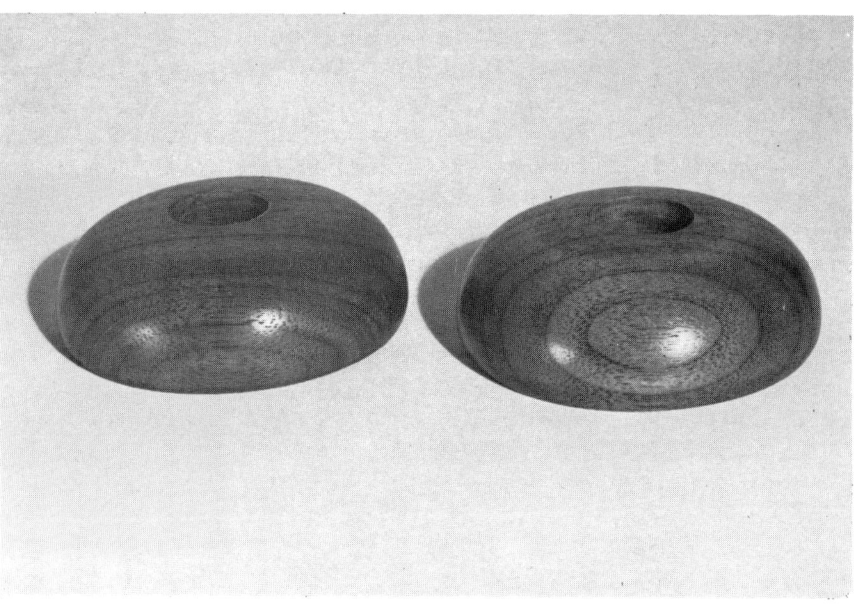

Illus.36. Candleholders

size of these candleholders, you may very well have a sufficient supply of scrap material available in your shop. Using a plastic template or school compass, pattern the blocks to a 3¼″ or 3½″ diameter. To avoid wasting wood, I vary the diameters of the blocks and turn the finished candleholders to the largest possible diameters. Of course, the type and size of the candles you plan to use in the candleholders should also determine the diameters of the blocks. But be careful not to make the candleholders too small or they will tip over during use. Now, prepare your blocks accordingly.

As with many of my projects, I tend to use black walnut, wild cherry, Osage orange (hedge), or one of the oaks. But I've also made the candleholders from some of the exotics such as bocote, purpleheart, or padouk. You may want to try some of the woods that are indigenous to your area. Don't forget your scrap box and your backyard woodpile. Since the project is so small, you may also want to consider using one of the exotics. This is a good opportunity to try some of the more expensive woods that you've been interested in turning. As with the salt and pepper shakers presented in Chapter 4, you may want to use the same wood you used for your dining-room table. Since these candleholders don't require very much wood, I hope you will feel free to do some experimenting with different types of wood.

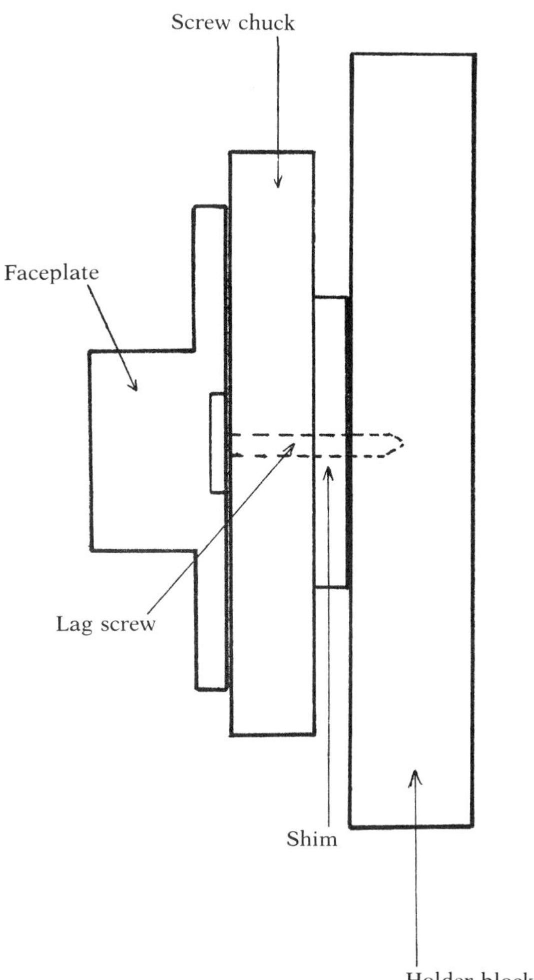

Illus.37. Holder block and shim on screw center

2. If you want to turn the blocks using double-face tape, refer to the discussion on its use in Chapter 1. My own preference is to use the small, homemade screw chuck that is also discussed in Chapter 1. The chuck is a 1″ (4/4's)-thick piece of hard maple or similar wood secured to a 3″ faceplate. A ¼″-diameter lag screw penetrates the middle of the chuck to secure the turning block. For specifics on making this chuck, refer to Chapter 1.

To turn the blocks on the screw chuck, you must drill undersized holes partially through the blocks. With a ¼″-diameter lag screw in the chuck, I usually drill the mounting holes with a 3⁄16″-diameter or smaller high-speed steel bit. Examine the blocks and select the best surface for the top. Find or approximate the middle of the bottom of the block and then mark it with a pencil. Using the appropriate-size bit, drill at least ⅝″ into the block at the middle mark. If your turning blocks are thicker than 1″ (4/4's), drill the hole deeper to assure a good hold on the lag screw.

Mount the screw chuck on the lathe. Thread a block on the lag screw to see if it will be secure against the face of the chuck. If it won't be secure, you will need to make a shim. How far the lag screw extends from the face of the chuck will determine how thick the shim should be. You need to make the shim so that it will thread onto the lag screw. To make a shim, I cut a small 1½″ square of 1″ (4/4's) thick oak or similar wood and drill a hole through the middle using the same bit I used on the blocks. Before cutting the shim to an approximate thickness, thread a candleholder block on the lag as far as it will go and measure the remaining distance between its bottom and the surface of the chuck. Add at least ⅛″ to this distance and cut the shim to this width on the band saw. Thread the shim on the lag and then screw on a candleholder block to test for fit. The block should now be secure for turning (Illus. 37).

If you want a little extra security while turning on a screw chuck, place a revolving center in the tailstock and bring it forward so that it's secure against the block. Let the revolving center penetrate the surface of the block. You can use this centered indentation as the mark for drilling the hole for the candle.

Using a shim between the chuck's surface and the candleholder block makes turning easier because it eliminates the possibility of cutting into the chuck with your tool when you are turning the block. Normally the small tenon that is cut into the chuck's surface will prevent your damaging the chuck with your tool. The shim procedure simply gives you even more room for turning the block.

3. It's best to turn the block using a ½″ or 1″ roundnose scraper with the lathe running at about 2500 rpm. After the block has been turned true, realign the tool rest so that it's facing the front edge of the block. Using a scraper, shape the block by removing the edge and rounding the surface. Continue rounding the surface until you have shaped the block as desired. After removing the tailstock, align the tool rest in front of the block and lightly clean up the top surface. Leave the indentation made by the revolving center; this is where you will drill the candle hole. Don't use too much tool pressure when you are working on the top surface because the lag screw could slip in the block.

Now, realign the tool rest to the side of the block and, with a ½″ square-nose scraper, chamfer the bottom edge of the block. You only need to make the chamfer about ⅛″ wide. The chamfer tends to give the candleholder a more finished appearance.

4. Sand the turned candleholder with 100-grit abrasive paper; then follow this grit with 150- and 220-grit. If your tools were sharp, you should only need to do a minimum of sanding.

5. The next task is to drill the hole in the candleholder for the base of the candle. However, there seems to be a lack of uniformity in the base diameters of candles. More often than not, the candle base is either too large or too small for the hole in the candleholder. I've heard that a 13/16″-diameter base will accommodate most candles; but over the years, I've used a 13/16″-diameter Foerstner bit to drill the base holes and I still encounter candles that don't fit. I also use a ⅞″ multispur bit on occasion. Either diameter should do. However, if you plan on using a particular kind of candle, check its diameter at the base and then drill the hole accordingly.

Although I prefer the Foerstner and multispur bits for drilling the candle-base hole, the various power bores and spade bits will also do the job. The type of bit you use should be at least partially determined by how you plan to drill the holes in the turned blocks.

If you used blocks that were thicker than 1″ (4/4's), you can probably drill the candle-base holes using the tailstock, a Jacob's chuck, and a bit. You should make the base hole at least ⅝″ deep if possible. The problem with drilling the 1″ (4/4's)-thick blocks that were turned on the screw chuck is that the lag screw penetrates too far into the block. Thus, drilling the holes on the lathe, in most instances, is not possible. However, if you turned your blocks using double-face tape, you can drill them on the lathe and as deep as you wish.

Since I turn the blocks using the screw center, I drill the candle-base holes using a drill press. The turned candleholder can't be clamped during the drilling procedure; so I wear a leather glove and hold the candleholder in place. Be certain you have a sharp bit or the candleholder may spin from your hand.

You can fill the lag-screw hole in the bottom of the candleholder with either wood filler or a plug (see the plug procedure discussed in Chapter 5). After you've finished the candleholders (see Chapter 43 for finishing products and procedures), you can cover the bottoms with felt so that they won't scratch the surface of the table. Another procedure involves using four felt dots: Place three of the dots around the bottom edge and the final dot over the lag-screw hole.

7
PINE PLATE—PLAIN OR INLAID

In recent years I've noticed that wooden plates are becoming more and more popular for both serving pieces and wall hangings. Wooden plates seem to be very much a part of the current interest in things with a country theme. Most of the wooden plates I've seen have been made from basswood on high-speed production equipment. However, a few years ago I decided to make some plates from pine.

Although I've turned pine plates in all diameters, my favorite plates tend to be the larger ones. These are cut from standard No. 3 common 1″ × 12″ pine, which is available at most local lumber dealers. If you've never turned pine, you're in for a pleasant surprise. While your tools need to be sharp, pine is relatively problem-free when it comes to turning. And the mouse that tends to develop at the cross-grain areas can be easily sanded to a smooth finish either on or off the lathe. You can leave these pine plates plain or decorate them with either tole paintings or inlays.

The large surface of the plate, turned from 1″ × 12″ pine, is a great area for an inlay—either surface or sculptured. Surface inlays are those designs that, when in place, will be flush with the plate's surface. Sculptured inlays, on the other hand, extend out from the surface and are two-dimensional. As Illus. 38 indicates, inlay designs do not have to be very complex. The inlay process itself is not terribly difficult as long as you keep your design fairly simple. Because pine is soft, you can cut into it easily using either a small hobby knife or a Moto-Tool with miniature router bits. For a detailed discussion on the inlay process, you may want to refer to my book, *Making Small Wooden Boxes* (Sterling, 1986). More about doing inlay work on turned plates on page 51.

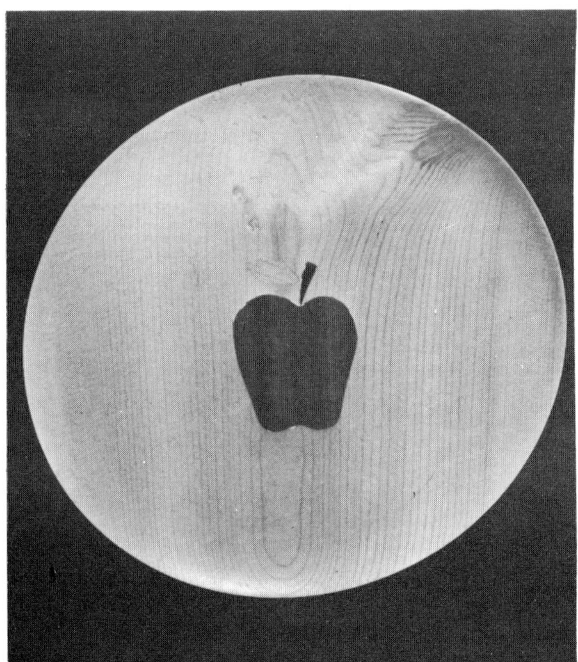

Illus.38. Turned inlaid plate

To simplify the preparation of blocks for turning, I use a circle-cutting jig on the band saw. The jig is homemade and looks rather crude (Illus. 39), but it's extremely effective for cutting perfect circles of wood. I have it set up so that I can make circles with diameters ranging from 2″ to 16″. You may want to consider making a jig yourself. It has many other applications in addition to preparing large pine circles for turning. Although you can pattern a block with a school compass, it's sometimes difficult to cut a perfect circle freehand. You will find when you're turning the large-diameter plate blocks that the more perfect the circles, the easier and safer they are to turn. More about this circle-cutting jig as we proceed with the project.

As a rule, I use glue blocks or discs to secure the plate block to a faceplate. Because of its large diameter, the outer edge of the plate moves extremely fast when it's on the lathe. But with the glue-block method, you have some assurance that the block won't fly off during turning. The nature of this project prevents you from securing the plate block directly to the faceplate with screws. However, you may want to explore some other methods for securing the plate block for faceplate turning.

TASKS:

1. As indicated, I make the larger plates using No. 3 common 1″ × 12″ pine. The actual dimensions of 1″ × 12″ stock is ¾″ × 11¼″. You need to be aware of these dimensions as you plan for the project. If this will be the first plate you've turned, you may prefer to start with a plate that has a smaller diameter and thus cut a smaller block from 1″ × 6″ or 1″ × 8″ pine. A smaller initial project can be a more manageable and economical way to gain some experience. I frequently use pine when I'm developing a new turning design or project. It's an excellent wood for turning in addition to being both available and affordable.

To make the pine block for the project, you can use a school compass and the band saw, and cut freehand, but my own preference is to use the circle-cutting jig, shown in Illus. 39. Although the jig looks crude, it's actually very effective for cutting perfect circles of wood to almost any diameter on the band saw. In addition to using it for cutting pine blocks for this

Illus.39. Circle-cutting jig

project, you can also use it for cutting hardwood circles.

I make the circle-cutting jig from a piece of 1″ (4/4's)-thick hard maple, but you may prefer using a different hardwood. The strip that holds the jig on the band-saw table is made from pine and it's the exact dimensions of the table's mitre groove. Refer to Illus. 40 for the approximate dimensions and the layout of the jig. With this jig, you can cut circles that range from 2″ in diameter on up through 16″ in diameter, and you can cut them in increments of 1″. You can also design the jig to cut in either larger or smaller increments, if desired.

Following Illus. 40, prepare the base section of the jig and cut out the one corner to the suggested dimensions. To secure the mitre-groove strip at the correct location, place the jig base on the band-saw table. Refer to Illus. 40 to see how you should place the board on the table in relation to the blade. Once the base is properly aligned in relation to the blade, mark the width of the mitre groove on the front edge of the jig base. Now remove the jig base from the saw table.

You need to transfer the mitre-groove markings onto the bottom surface of the jig using a large square. Transfer the marks on the front edge to make two parallel lines running the full length of the jig's bottom surface. Secure the pine strip, cut to this same width, between the two parallel lines. When it's secured in place, this strip will move in the mitre groove and guide the jig as you make the cut.

Referring again to Illus. 40, note the line that runs across the top surface of the board where the various holes are indicated. Using the suggested dimensions, make a similar line on the top surface of your jig. Before marking along the line for the various diameters, it's important to remember that, on the jig, you're actually measuring only one-half the diameter of the circle. It is really the radius that you're laying out on the jig's surface. For example, the first mark that is used for making a 2″-diameter circle is, in fact, only 1″ away from the band-saw blade, and the point for making a 3″-diameter circle is 1½″ from the blade. Once you understand this principle, it will be easy for you to make your jig-cut circles to any diameter you wish.

Now place the jig on the table with the pine strip in the mitre trench. Align the jig so that the line across the top surface is in perfect alignment with the front edge of the blade. Place a ruler against the side of the blade and let it extend along the line on the surface of the jig. This procedure will allow you to mark the various points on the line for cutting the range of diameters. As you will discover shortly, a small hole is drilled at each marked-diameter point. Mark out the diameters you want on the line. If you make your jig the same width as mine, you should have a range of diameters from 2″ to 16″.

Using a ⅛″-diameter high-speed bit, drill a hole through the jig at each marked-diameter point. I use a drill press to make certain the holes are drilled straight. Turn the board over and countersink each hole in the bottom surface. Use a ⅞″-long wood screw as a pivot and make sure it penetrates the board from the bottom. The top surface of the screw head must be flush with the bottom surface of the jig or it will not rest properly on the surface of the saw table. Countersinking each hole allows the screw head to be flush with the jig's bottom surface.

The pivot-screw should have a diameter of at least ⅛″ so that it doesn't flop inside the jig's diameter holes. It should extend above the top surface of the jig by no more than ⅛″. I sharpen the point of the screw with a file so that it will penetrate the pine circle to be cut. Depending upon the diameter of the circle to be cut, move the pivot screw to the appropriate hole (Illus. 41).

To use the jig, cut a board to square dimensions, from which you can cut the desired diameter. Find the middle of the square board and make a slight indentation with a nail. Set the jig's pivot screw to the desired diameter and place the jig on the edge of the band-saw table. Place the board's middle indentation on the pivot-screw point. Turn on the band saw, using a ¼″-wide blade. Move the jig into the blade until the surface line on the jig is in alignment with the front edge of the blade. You will be cutting into the board during this alignment procedure.

Once the line and the blade are in alignment, rotate the board through the blade and you

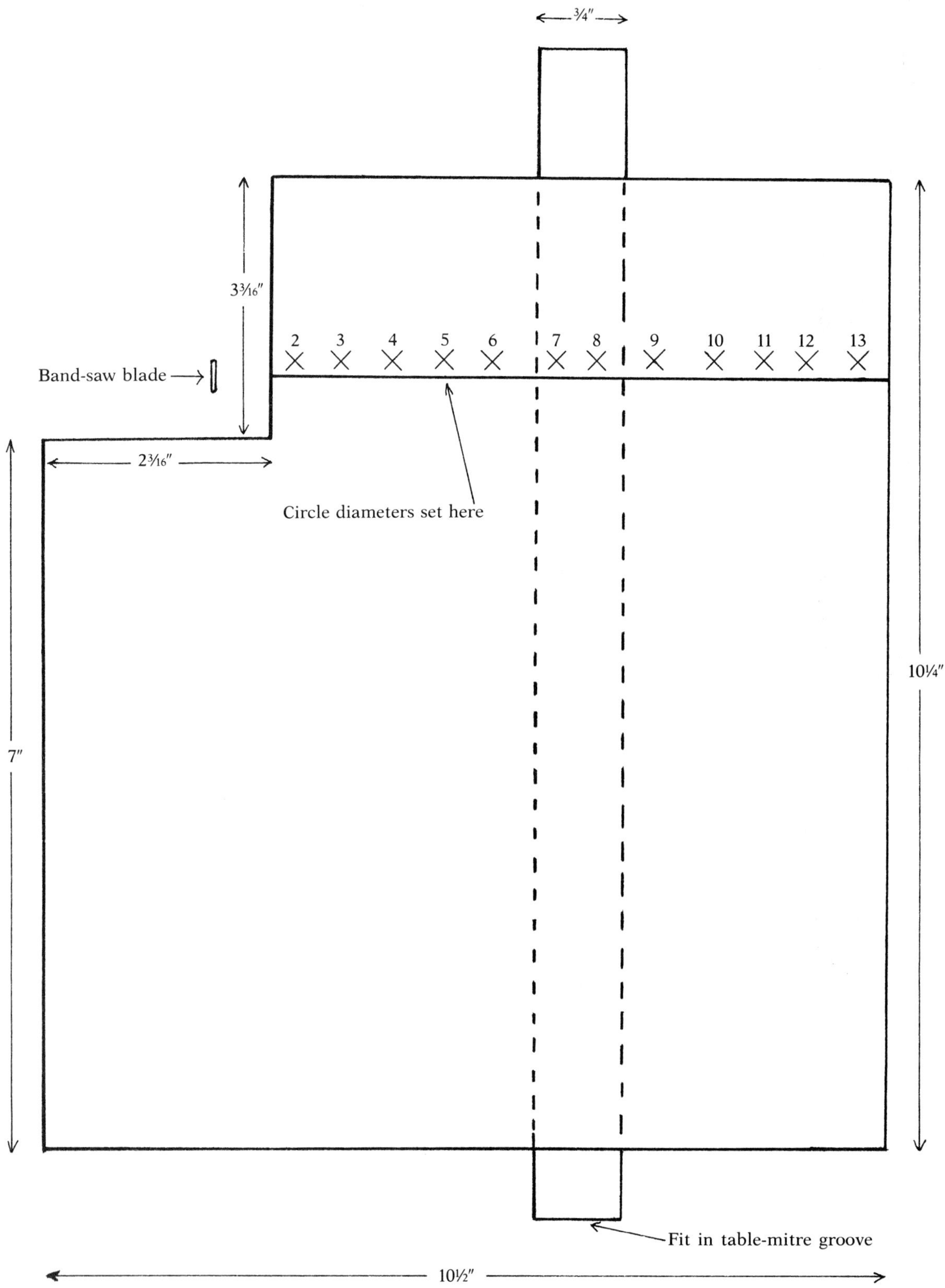

Illus.40. Circle-cutting jig

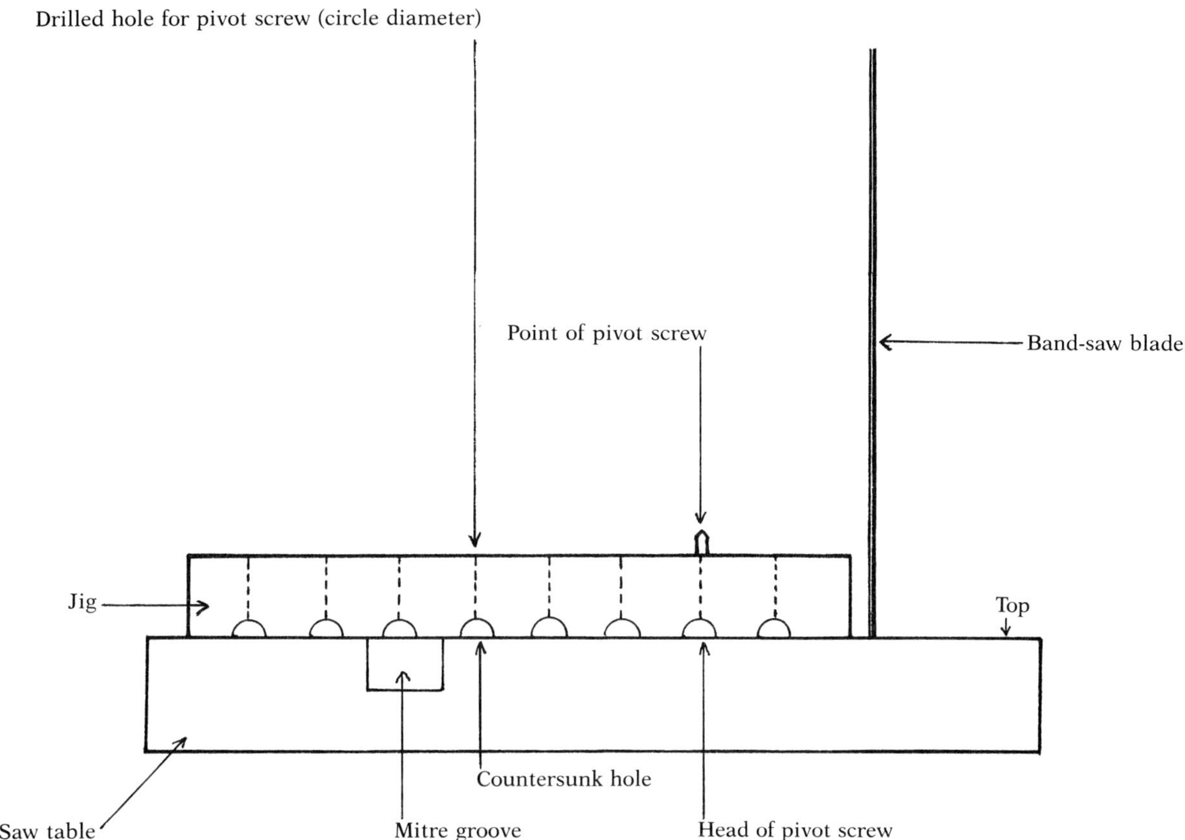

Illus.41. Back edge of circle-cutting jig

should get a perfect circle. You may have to make a few practice cuts to get the feel of the procedure. The important thing is to have the surface lines on the jig and the front edge of the blade in alignment. The board being cut covers a portion of the line, but you can still see part of it for the alignment procedure. A circle that is cut perfectly round on this jig is considerably easier and safer to turn than one cut freehand. Also, the jig can be very helpful with a range of other projects.

To make the turning block for the project, cut a piece of 1″ × 12″ stock to a square (11¼″ × 11¼″). Find the middle of the square by making diagonal lines from the four corners. Where the lines intersect is the middle. Using a nail, make an indentation at the midpoint. The indentation needs to be deep enough so that the block will fit over the pivot screw on the jig and lie flat on the surface of the jig. Set the jig's pivot screw to cut an 11″-diameter circle. Place the board's indentation on the pivot screw, slide the jig forward into the moving blade, and then cut the circle.

2. As indicated earlier, I turn the plate using a 3½″-diameter pine glue block that is attached to a 3″ faceplate. To prevent the block from possibly flying off, I glue the disc directly to the bottom surface of the pine circle. For perfect centering of the glue disc on the back surface of the plate, place the point of a school compass in the indentation made earlier. Pattern a circle that is 3½″ in diameter. Spread white glue on the glue disc and place it within the circle line. Clamp the assembly until the glue is dry. I use small bar clamps for this procedure. You may have to place a board over the glue disc in order to properly clamp it. Be certain the glue disc doesn't slide in the glue while you're clamping it. Allow ample time for the glue to dry. Attach a 3″ faceplate to the secured glue disc.

3. To turn the plate, I run my lathe at 1475 rpm. This is slower than my normal turning speed, but it's preferable for projects with large diameters. The edge of the plate may travel too fast if you turn at the higher speeds. Should you want some added support and security, place a revolving center in the tailstock and bring it forward into the surface of the plate.

I use my usual 1″ roundnose scraper to true the edge of the plate and to do the initial shaping. I roll and taper the edge down the back surface of the plate. As a rule, I finish turning the edge of the plate before I start on the front surface. You should use a freshly ground scraper with a good wire edge when you are turning pine. A well-ground tool will greatly enhance the surface of the project, especially when you're working with pine.

Realign the tool rest for turning the front surface. If your tailstock is in place, you can use a 4″-long tool rest and still have space for turning. Using a 1″ roundnose scraper, I begin removing wood near the middle and then move to the edge. The inside edge should match the contour of the outside edge. Usually, I try to leave a uniform thickness of about ⅜″. The exception is the very edge of the plate. Both surfaces taper together at the edge. With the scraper, I slightly round the edge to prevent any possible chipping of the plate during use. You can further clean up the edge with abrasive paper.

If you're using the tailstock for support, remove it and slide the assembly out of the way. Place a 12″ tool rest in the support and finish turning the middle area. You want the plate surface to be flat; so spend a little extra time with your tool. Usually you can remove the middle dimple by passing over it very lightly with the roundnose scraper. If you plan to use an inlay, you will remove the dimple during the inlay procedures.

4. When you have completed the turning, the plate probably will need some sanding. Pine is rather coarse-grained and the cross-grain areas tend to be somewhat rough or mousey in appearance. To avoid scratching the surface, I generally begin sanding with 150-grit abrasive paper. I use a piece of carpet pad as a sanding block. This not only absorbs the friction, but it also ensures more uniform pressure while you're sanding. If you were unable to remove the middle dimple with a tool, you can easily do so now with abrasives.

After going over the edge and the front surface with 150-grit paper, I move up to 220- or 240-grit. For a final going over, I use (0000) steel wool, which smoothes the surface and brings life to the wood. If you decide to use steel wool, be sure to blow or wipe off all the small wire strands prior to finishing.

5. There are a number of options for removing the glue block from the base of the turned plate. One effective way is to use a parting tool while the assembly is still on the lathe. You will need a parting tool with a relatively long shank to reach from the tool rest to the block. Place the parting tool along the bottom surface of the plate and into the glue block. Make the entry cut with the parting tool at the glue line. This helps to remove some of the mess that always accompanies the use of glue blocks. Unless you're good at catching a revolving plate, don't entirely separate the block and the plate; instead, leave about a ¼″-diameter tenon between the two. Shut off the lathe and, using a handsaw, cut through the tenon. With a handsaw, you can place the tenon right next to the plate's bottom surface. Be careful that the plate doesn't drop onto the bed of the lathe because the edge of the plate can chip quite easily.

Sand the bottom surface of the plate using a large belt sander with a 100-grit (or finer) abrasive belt. Follow this with a range of abrasive grits, sanding by hand or with a small electric finishing sander. Be certain you sand with the grain. You may want to rub the surface down with (0000) steel wool.

Another way to remove the glue block from the plate is to use an old hand-plane blade or a wood chisel. Do not try to separate the block at the glue line. If you do this, pieces of the plate bottom can come off with the glue block. The best procedure is to rest the edge of the glue block on a support surface and chisel away parts of the block. After completing this procedure, sand whatever remains using a large belt sander and a coarse belt. You can then finish-sand the surface with a range of finer abrasives and a pad sander.

Incidentally, I wouldn't recommend using

the band saw to remove the block. Either the band-saw blade will grab the glue block during the initial cut or you will chip the edge of the plate during the cutting procedure. Also, most band saws will not accommodate the diameters of large plates.

If you want the plate to remain plain, it should now be ready for finishing. For finishing products and procedures, refer to Chapter 43. If the plate is to be tole-painted, it's best to discuss finishing with the artist.

6. As Illus. 38 indicates, you can do a surface inlay on the turned plate. I'm especially fond of inlay work; so I tend to include it on many of my projects. You can use veneers for surface inlays, but I prefer cutting the material from solid stock. I generally rip the stock on the band saw to a thickness of at least 1/16". You may want to use a 1/2"-wide band-saw blade with at least 6 TPI (teeth per inch) to cut the material for the inlay. This is a good general-purpose blade and is especially effective for cutting inlay material.

The initial task in inlay work is developing an acceptable and manageable design. A piece of fruit—an apple or a pear, for example—is a good first design. A bouquet of flowers with stems and leaves is another good first design. You can, of course, make an inlay design of almost anything. The real issue is whether or not you're willing to spend the time and energy.

For purposes of demonstrating the surface-inlay process, I will use the apple design, shown in Illus. 38. I made the apple from padouk (red), the stem from ebony (black), and the leaf from poplar (green). The apple is 3" high and 2 1/2" wide. The dimensions of the stem and leaf are up to you, but they should be proportional to the apple.

Using a piece of padouk that is at least 3 1/4" wide, saw off a slice that is approximately 1/16" thick. Using smaller pieces of ebony and poplar, cut slices of each for the stem and leaf. You will be placing the sawed surfaces of the various parts inside the recessed areas on the plate surface. This way, they don't have to be sanded. Also, the sawed surfaces are good for gluing.

You may want to make a paper pattern for the apple, but I usually just draw the outline directly onto the surface of the padouk. Play around with the shape of the apple until you are satisfied with it. You will want the grain of the wood running up and down on all the inlay parts. This looks better and, to some degree, makes the parts easier to cut. Now pattern the leaf and stem on their respective woods.

If you have a large scroll saw, cutting the inlay parts should be fairly simple. I use my old 24" Rockwell scroll saw to cut the various parts, but you may have a different type of tool that works just as well. You can also use a small hobby knife with a new pointed blade (Illus. 42). Cut the various pieces along their pattern lines. When you are cutting, be careful that the parts don't split along the wood grain. When you're cutting across the grain, rock the knife and its point so that it will cut the wood fibres. *Take your time with this procedure and be very careful with the knife. The blades are almost surgically sharp and can make a serious cut very quickly.*

Illus.42. Hobby knife (photo courtesy of Stanley Tools)

When you have cut the parts, lightly sand their edges using 220-grit abrasive paper. Don't round the edges; just clean them up, removing any splinters or fibres that may remain. You want a good square edge on the inlay parts to ensure a neat fit in their respective recesses. You also should handle the parts with care since they can split rather easily.

Place the apple in the approximate center of the plate. The wood grain of the plate should be running up and down. Using either a sharp pencil or a scribe, trace the apple outline onto the plate's surface. This will be your cutting line; so make the line visible. Repeat this tracing process with both the stem and the leaf. When I'm using a pointed scribe, I sometimes sprinkle blue chalk dust into the scribed lines. This helps you see the traced lines when you are

Illus.43. Dremel Moto-Tool with router attachment (photo courtesy of Dremel Tools)

cutting the recess. While you can purchase powdered chalk, I generally use the small pieces of chalk that come with a child's blackboard.

The recesses in the plate's surface should ideally be 1/16″ deep. Unless you are routing the area with a Moto-Tool using a router base and microbit, it's almost impossible to achieve this exact depth throughout the recesses (Illus. 43). When you're using a knife, you do need to be careful so that you don't exceed the 1/16″ depth. You want the inlay to be flush with the plate's surface when it's in place. Thus, the depth of the recess is critical.

The initial cuts into the plate's surface should be along the pattern lines of all three pieces. I cut on the inside edge of the lines. This usually results in a good fit. With a pointed blade in the knife (Illus. 42), you can cut out the recessed areas for both the stem and the leaf. Make the cuts at an angle, slanting across to the previously made pattern-line cuts.

To remove wood from inside the apple recess, I use a router blade on the hobby knife. With this blade you can make scooping cuts while also controlling their depth. It's best to cut with the grain and up to the margin incision that's already been made. Since pine cuts and splinters very easily, take your time. Should you cut certain areas of the recess too deep by accident, fill them in with some wood filler; then sand the filler flush when dry.

When you have prepared all the recesses, test the various inlay parts for fit. If an inlay piece is too large, widen the area accordingly. Test and cut until all the pieces fit neatly into their respective recesses.

Spread a thin coat of wood glue in the recesses, being certain you get the glue next to the edges. You want the edges of the inlay parts to be firmly glued in place. Now set the inlay parts into the glued recesses and press. Then wipe off any excess glue that may squeeze out along the edges. It's difficult to clamp this area; so you will need to weigh it down with something. I frequently place a sheet of wax paper over the inlay and the surface of the plate; then I place a board over the wax paper and, on top of the board, a coffee can filled with nuts and bolts or sand or anything else that's heavy. The wax paper prevents the board from getting glued to the inlay and the surface of the plate.

When the glue is dry, clean up the surfaces of the inlay and the plate with 220-grit abrasive paper. If you have some gaps between the inlays and their recess edges, fill them in with a paste made of pine sawdust and white glue. When the paste is dry, sand the area carefully with the grain. Now turn to Chapter 43 for some ideas on how to finish the plate.

8
ICEBREAKERS

This project is not a verbal tidbit used at social gatherings, but rather a functional tool that's used in a kitchen or bar for smashing ice. This is one of those projects that is both fun and relatively easy to make, and it gives you the chance to do some spindle-turning. If you want

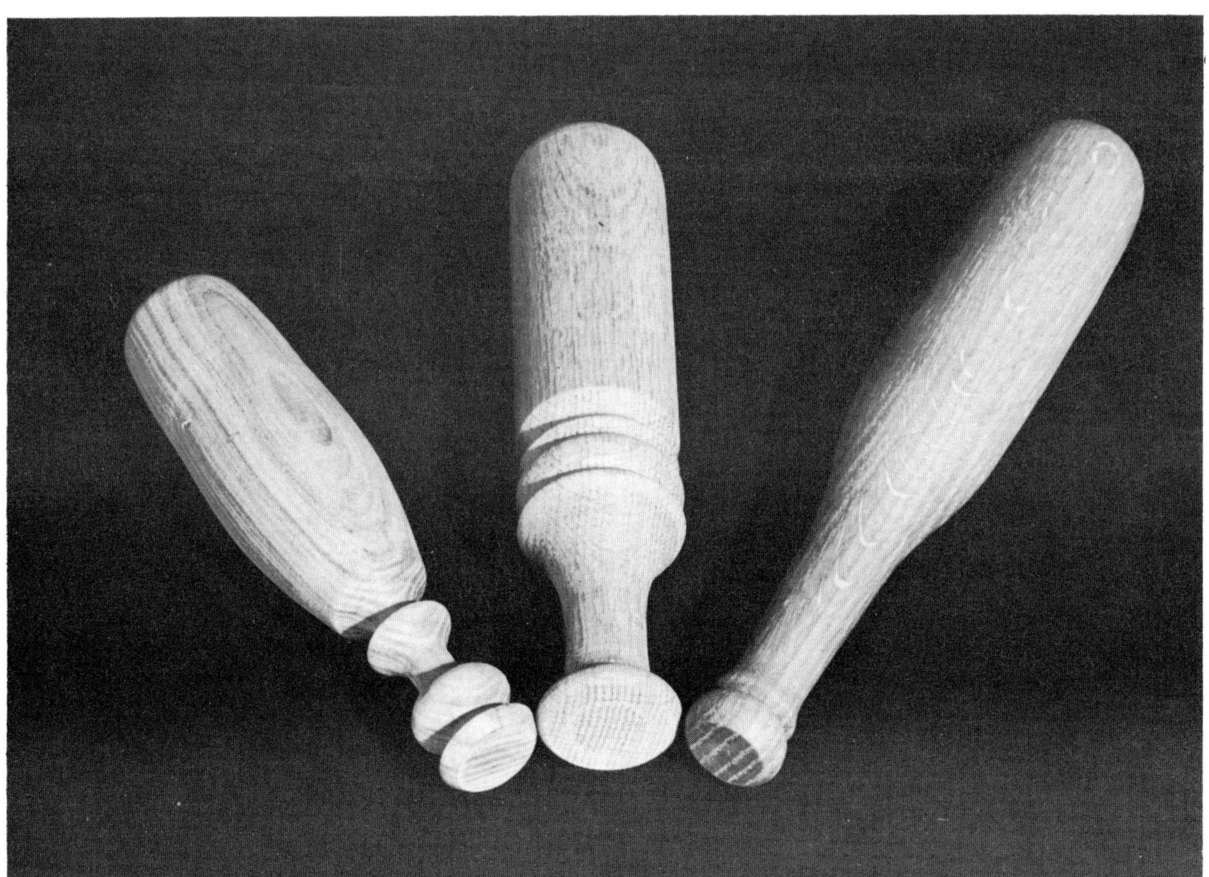

Illus.44. Icebreakers

Illus.45. Roughing-out breaker blanks with 1¼" roughing-out gouge

to make items in quantity for small gifts, you may want to consider this project.

As you can see in Illus. 44, icebreakers resemble heavy-ended baseball bats. I usually turn icebreakers from scraps of oak, hard maple, beech, Osage orange (hedge), ash, pecan, or hickory—or any other woods that I have in my shop. Since the project does not require very much wood, often you can use scrap material that you have around your shop. For an unusual-looking icebreaker, you may want to laminate contrasting woods or even use a piece of veneer for an inlay, but be sure to use a waterproof glue. Icebreakers are usually subjected to a lot of water and regular wood glue will break down over time.

You also might want to place some decorative plugs at various critical points on the icebreaker. You can cut your own plugs using a plug cutter or you can use pieces of dowel instead. Drill holes of the appropriate diameter using a brad-point drill and then glue the plugs in place. If you place the plugs in what will be the handle, you will need to set them in quite deep so that they won't be turned away. The finished diameter of the handle is smaller than that of the head; so you must place the plugs accordingly.

As with most other projects, icebreakers can be greatly enhanced and personalized if you use a bit of thought and ingenuity. Much of the fun in a project involves drawing on your own ideas and techniques. With some reflection and a bit of design work, you can make a unique icebreaker.

TASKS:

1. Prepare a turning blank that is at least 1¼" (6/4's) square and 12" long. Depending upon your own design considerations, you can either reduce or increase these dimensions. You should also consider the availability and size of the scrap pieces that you may want to use. Pre-

pare the blank so that the grain will run lengthwise, or parallel with the lathe bed. If you want to prepare a glued-up spindle, use 1″ (4/4's)-thick stock. This will give you a larger diameter to work with and more weight for smashing ice cubes. As previously suggested, you may want to use contrasting woods or a strip of veneer between similar boards, but remember to use a waterproof glue for any laminating work.

If you plan to use inlaid plugs, this is the time to cut them and insert them into the blank. You may want to consider placing a centered plug in both ends of the spindle. Remember that if you place any plugs in the handle area, you should set them deeply into the wood. When you're placing plugs in a square blank, you need to set them in so that they won't be turned away as you round and reduce the blank's diameter. It's a good idea to use waterproof glue on the plugs. If you use regular wood glue, in time water could loosen the glue bond.

2. For spindle-turning the blank, I use a four-pronged drive center and a pointed revolving center in the tailstock. Before mounting the spindle, I make diagonal lines from the four corners on both ends. This greatly simplifies centering the drive prong and the revolving center. Using an extra-pronged drive center and a wooden mallet, I make the drive indentations in one end of the blank. A good procedure is to set one end on the floor, align the drive-center prongs on the lines, and then tap it with the mallet. This procedure ensures a good hole on the drive when the blank is mounted. It also eliminates the need to put unneeded pressure on the bearings when the blank is driven into the prongs while it's mounted on the lathe. Incidentally, be certain that the center point of the four-pronged drive is sufficiently extended. If it's not, use a hex wrench to loosen it in the prong and then pull it out. Be sure to tighten the hex screw again after the point has been extended.

Mount the blank between centers. I move the revolving center in so that its point penetrates well into the marked center of the blank. Be certain the blank is centered at the point where

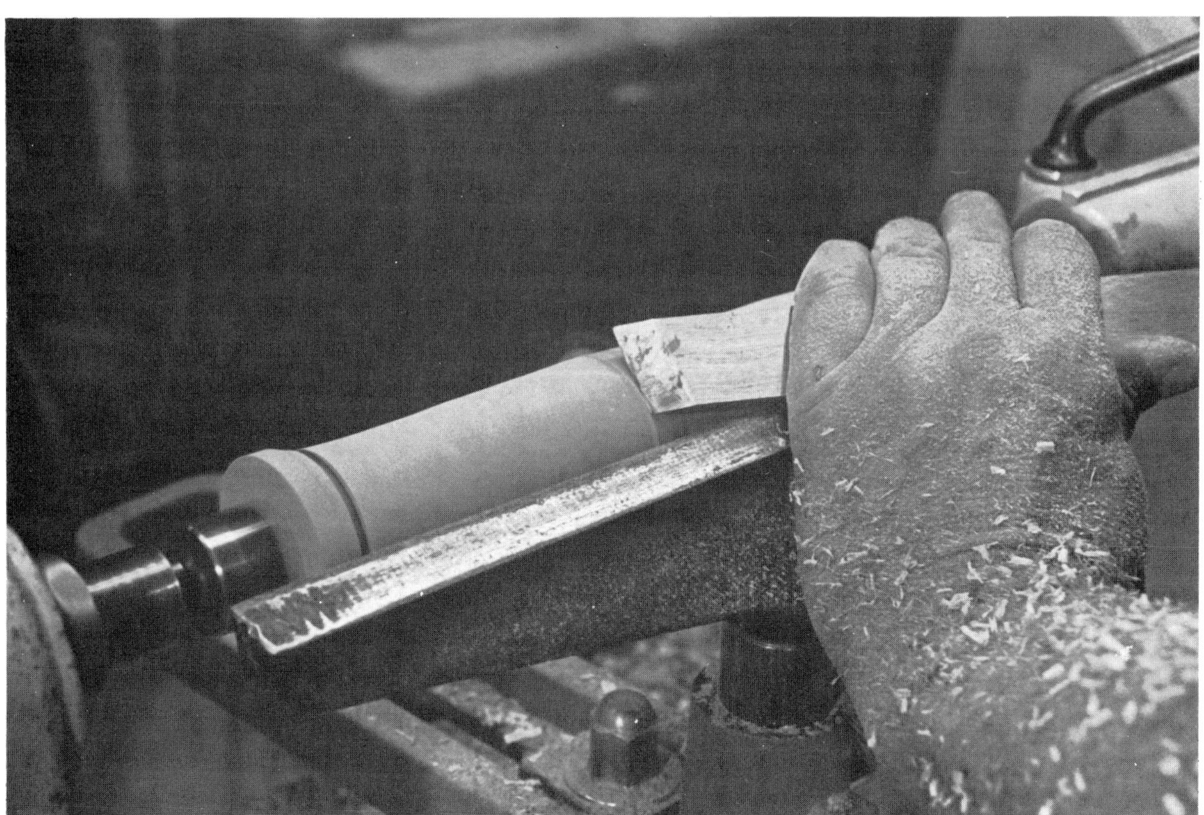

Illus.46. Planing icebreaker with 1″ skew chisel

Illus.47. Rounding breaker head with 1″ skew chisel

the diagonal lines intersect. To turn the icebreaker, I run my lathe at 2500 rpm. Use your tool of choice for roughing-down the blank—for me, a 1¼″ roughing-out gouge is ideal. It's quick and leaves a reasonable surface, and it's also a lot of fun to use. Be sure to ride the bevel, once the edges are removed (Illus. 45). If you would rather not use a gouge to true up the blank, you can use a 1″ roundnose scraper instead.

You will want to retain as much wood mass as possible at the head; so plan your handle area. Since you can actually hold and use the icebreaker with just a few fingers, an area that is approximately 4″ long is usually adequate for the handle. This area should also have a reduced and tapered diameter. You can turn a small bead or two that will separate the handle from the head section. A cove would also look good. You will definitely want to have a bead on the end of the handle because a bead will help to prevent the icebreaker from slipping out of your hand during use.

After I have rounded the spindle, I plane its surface and do some initial tapering using a 1″ skew (Illus. 46). However, you may want to use a shallow spindle gouge or a roundnose scraper to remove the wood more quickly. If you're planning to have a bead or beads between the head and the handle, plan your cuts accordingly. You should also plan to leave at least ½″ of the spindle between the drive center and the bottom edge of the handle. This ½″ drive tenon will make it easier for you to turn the bottom bead on the handle. It will also help in preventing you from hitting the drive prongs with your tool. The bead at the base of the handle should be approximately ¾″ in diameter, and the top of the head should be at least 1″ in diameter.

The top edge of the icebreaker should be rounded over so that it resembles the top edge of a baseball bat. You can readily do this with a ⅜″ gouge or a skew chisel. If you use a gouge, ride the bevel down the top surface. You want to leave a short tenon between the head and the revolving center. The gouge or chisel will cut the end grain and leave a clean surface on the top of the icebreaker. You can cut the supporting tenon to a diameter of ¼″ (Illus. 47). If you have a wood plug in the top surface, you will need to make the supporting tenon very short. You can saw off the remaining tenon after you have completed the spindle.

If you want to dress up the icebreaker with a

few more beads and coves, turn the spindle accordingly. An icebreaker is an excellent project for practising making beads.

3. If sanding is needed, start out with 150-grit abrasive paper and move on up through the finer grits. The type of wood you used for the icebreaker will partially determine which abrasives you use. You should keep the abrasive moving along the length of the spindle. Remember, the wood grain is running parallel with the lathe bed on this project. Sometimes I stop the lathe and lightly work on spots—for example, around the beads. I often use (0000) steel wool as a final rub on the surface. This tends to slightly burnish the surface and it also takes the abrasive scratches out of the wood. A good practice is to rub the piece, lengthwise, with the steel wool while the lathe is shut off.

4. Now remove the turning and band-saw the tenons off on both ends. You should be careful when you're band-sawing the tenons. Have a firm grip on the piece so that the blade doesn't pull it from your hands. Using a range of abrasive grits, sand both ends of the icebreaker.

If you want to be able to hang the icebreaker, thread a small eye screw into the base and attach a piece of rawhide or drill a small hole through the handle.

Finish the icebreaker with a nontoxic oil. For further specifics, refer to Chapter 43.

9
LOG WEED POTS

This is one of those projects where the wood can actually be obtained from your backyard woodpile. If you don't have a woodpile, you can often find small logs on lots being cleared for construction. Another good source for material is land that power companies are clearing for power lines. If all else fails, try to find a friend with a woodpile who will be willing to part with a few small logs, and, in return, promise your friend a hand-turned weed pot.

I often turn weed pots from logs that range from 3″ to 6″ in diameter, and from 4″ to 6″ in length. After you've turned a few small ones, you may want to try some larger logs. You will want to find logs that still have their bark tightly in place. This often means that the wood

Illus.48. Log weed pots

is not completely dry, but, for this project, that's ideal. The finished pot may develop some checks and cracks and even lose a little bark, but this only enhances the piece. If the wood or bark should eventually deteriorate too much, you'll have to discard the pot. However, you will have had the pleasure of turning it and also of enjoying it for a while, and there's no reason that everything we turn should last indefinitely.

Frequently the logs that are best suited for weed pots are not perfectly straight. Even though they are centered on the faceplate, they usually run off-center on the lathe. Thus, you will find yourself doing some off-center turning that, at first, may seem somewhat intimidating. As a matter of fact, weed pots actually look much better if they're turned off-center. When a log looks too straight, I deliberately secure it off-center in order to obtain the effect that I prefer. As you can see in Illus. 48, I only turn the top portion of the log. The extension from the top surface that holds the weeds looks much better when it's slightly off-center. As you do this project, you will discover ways to reduce the hazards of off-center turning and still obtain some very unusual effects.

TASKS:

1. Using logs that range from 3″ to 6″ in diameter, cut off sections that range from 4″ to 6″ in length. To cut the chunks, use a band saw with a ½″-wide blade. You should cut at least one end of the chunk square. This will simplify mounting the block on a faceplate, and it will also prevent the block from being too far off-center.

As discussed in previous chapters, there are any number of ways to mount the block onto a 3″ faceplate. If the diameter of the log is 3½″ or more, I usually secure the faceplate directly to the block with long screws. You need a diameter of at least this size or else the block will split when penetrated by the screws. Since you are using screws in end grain, be sure they're long ones. You want them to hold while you're turning, especially when the block is off-center. When I use screws, I save some sawdust from the piece being turned, mix it with white glue, and then use the paste to fill in the screw holes. This type of filler usually fills and conceals the screw holes quite well.

For logs with a diameter of less than 3½″, I use a pine glue block or disc. As discussed in Chapter 1, I cut the blocks to a diameter of at least 3½″. This diameter will usually prevent any splitting from the faceplate screws. Since you will be gluing end grain to the glue disc, don't use paper between the pine disc and the log. Glue the disc directly to the log using either white or yellow glue. You should use a little extra glue because the end grain tends to soak up glue rather quickly. If you use regular wood glue, be certain to clamp the glue disc and the log long enough for the glue to properly dry. It's important to have a good strong bond between the glue disc and the log. After the glue is dry, secure the faceplate to the glue-disc assembly.

2. When the assembly has been threaded on the lathe, you may want to bring the tailstock forward for both support and security. Use a pointed revolving center in the tailstock and force it into the top surface of the log. Before you begin turning, you may want to refer back to Illus. 48 to look at the area to be turned.

To turn the top area of the log, I use either a 1″ or ½″ roundnose scraper. You should grind the scraper, leaving a wire edge in place. You're cutting partially into end grain, so you want a good sharp tool. The tool rest should be aligned so that you can turn at the front edge of the log. Make the initial cuts into the upper side of the log, and work the scraper to the right, towards the tailstock. You're cutting a curved area that will result in a 1″-high pillar, or extension, that will hold the weeds. Thus, you want the top surface to have a slow curve that tapers to the top of the pillar. On smaller logs, I tend to turn the top of the pillar to a diameter of ⅜″. With this diameter, you will have enough wood for drilling a 3⁄16″-diameter hole in the pillar. On larger logs, I increase the diameters of the pillars and holes accordingly.

You may prefer to make a pot that has a shorter pillar with a larger diameter. Base your turning on the anticipated use of the pot, as well as on your own design preference. When turning the pillar, you may want to move the tool rest closer to it. You want good support for the scraper when you are turning the pillar. Tool clatter can, in some instances, break off the pillar. After you have turned the top area to your satisfaction, roll the upper edge of the side a little. This cut will remove some of the bark and expose the underlying wood, giving the piece a more finished appearance.

3. You will need to go over the turned area with some abrasive papers. Begin with 100-grit and move on up through some of the finer grits until you obtain the desired surface. When you have completed the sanding, cut the glue disc (assuming you've used one) from the log with a band saw. I cut into the log just above the glue line. This way, you can prevent having any glue residue on the base of the pot.

The base can be sanded on a belt sander and then finished by hand or with an electric finishing sander. Using the belt sander, I slightly roll the bottom edge of the log. As with the top edge, this procedure removes some of the bark and exposes the underlying wood. It also gives the entire piece a more finished appearance.

4. As I've mentioned, on logs with smaller diameters, I drill a 3⁄16"-diameter hole into the extension and down into the log. The indentation made by the revolving center is the point of entry for the drill bit. For a deeper hole on longer logs, I use an aircraft bit, which is significantly longer than a standard high-speed bit and available in a range of diameters.

After you have drilled the hole, lightly sand the top edge of the extension by hand. Carefully roll the edge with a fine-grit abrasive paper. You may want to refer to Chapter 43 for information on the finishing process.

10
SMALL WEED POTS

The small weed pots in Illus. 49 are but a few examples of the many designs you can turn.

Although the pots have drilled holes in them to hold some small stems, they can be decorative pieces in themselves. I've noticed that many people collect very small turnings. Everyone seems to have shelves or cabinets where they display small decorative items. In addition to small turned pots, many people collect small boxes. (You can find instructions for

Illus.49. Small weed pots

turning various small boxes in later chapters.) I've noticed that the weed pots and boxes are especially treasured if they're turned from unusual woods. A number of the pots in Illus. 49 were turned from spalted hard maple. People seem to be fascinated with the spalts, especially when they're used for smaller pieces.

In addition to being an enjoyable project to turn, a small weed pot can also be made from scrap material. I tend to save all my scraps, especially from thicker stock. I seem to have boxes of scraps all over my shop. While a great consumer of space, the scraps are truly useful when I want to turn a small project, such as a small weed pot.

If you are not inclined towards spindle turning, you can also faceplate-turn these small pots using double-face tape. You may want to reverse the grain of the block for faceplate work, but it's not mandatory. Glue blocks are also a good option with these small turnings. The point is that, with these small projects, you can choose from a range of options.

To demonstrate this project, I will present two different approaches: turning between centers and using double-face tape for faceplate work. This will give you some options as well as an opportunity to try something different.

TASKS:

1. *When turning a weed pot using the faceplate method*, I cut the block across the wood grain. (For spindle turning, the block should be cut with the grain so that it runs parallel with the lathe bed.) Using either a plastic circle template or a school compass, pattern a block, across the grain, with an approximate diameter of 1¾". You may prefer using a larger or smaller block, depending on the size of the scraps you have available. I often use stock that is at least 2" (8/4's) thick, but this thickness is more a matter of preference than necessity. I like to make the weed pot at least 2" high.

As I've indicated, when turning these small blocks using the faceplate method, I generally use commercial double-face tape. You need to cut a 3½"-diameter pine or plywood disc, on which you'll place the tape. This diameter usually prevents the disc from splitting when penetrated by the faceplate screws. You should use commercial double-face tape that is available from a number of suppliers. The double-face carpet tape that you find at hardware dealers won't work. It simply does not provide the bond that's necessary for holding the block in place while you're turning.

I use two or three strips of double-face tape, depending upon the diameter of the block to be turned. The length of the strips should also be determined by the block diameter. Given the cost of the tape, I use only the minimum required for a given project. As I mentioned in Chapter 1, I've had little success in reusing the tape. It's great for one use only; after that, its bond is unpredictable.

Cut strips of tape to length and place them, centered, on the 3½" disc. Remove the covering from the tape strips and place the block at the approximate center. Force the block down onto the tape and then clamp. I use a bar or C-clamp for this procedure. A few minutes of firm clamping is all that is required. With the block secured on the tape, mount the faceplate on the disc and thread the assembly on the lathe.

If you are concerned about the holding power of the tape, place a revolving center in the tailstock and move it into the block. This will give you both security and added support while turning. You will need to turn a few projects with the tape in order to develop some confidence in its use. While it's possible to force the block off the tape while turning, you really have to work at it. When you try to remove the turned piece from the tape, you will see how strong the bond actually is (Illus. 50).

To faceplate-turn the weed pot, I use ¼" and ½" roundnose scrapers. I true the block and do some initial shaping with the ½" scraper. The ¼" is ideal for final shaping, especially around the neck area of the pot. With the ¼" scraper, you can also shape the bottom edge without any fear of knocking the block off the tape. Be sure your tools are sharp when you are cutting these smaller blocks on tape. With sharp tools you don't need to use so much pressure.

After you have shaped the side, realign the tool rest and clean up the top surface of the block. You will want to remove the tailstock for this procedure. A good tool for cleaning up the top surface is a ½" square-nose scraper. Use a light touch when you apply the scraper. If possible, leave the indentation made by the revolving center. This marks the exact midpoint of the block, and you can use it for drilling the hole into the turned pot later. Use some abrasive paper, depending upon the condition of the surfaces. You may want to start with 100-grit and move up to 220-grit. As a final procedure, I often go over the surfaces with (0000) steel wool.

For spindle-turning the weed pot, prepare a 2″-square blank. You may want to increase the length from 2″ if you prefer more height to the finished pot. Added length will also allow you to leave a short tenon on both ends of the blank. These tenons will hold the drive and revolving centers. They can be band-sawed or one-cut with the parting tool after you've turned the piece. The tenons or stubs often make turning small blanks between centers much easier.

As I've mentioned, the overall dimensions of the project are a matter of personal preference. Very often, my designs—especially on smaller pieces—will be determined by available stock. I'll vary the length or diameter of a piece according to what happens to be in my scrap box. Even if I do some paper-and-pencil design work for a small project, I usually will alter the design during the actual turning process. The lathe is endlessly versatile and will often allow a shape to emerge that is more attractive than any planned design.

When spindle-turning small blanks, you have a number of options on the drive end or headstock. I generally use a ball-bearing pointed revolving center in the tailstock. One drive-center option is to make a two-pronged drive from a ¼″-diameter straight router bit. This bit is described in Chapter 2. To turn with the pronged bit, you need to mount it in a Jacob's chuck that is held in the headstock by a Morse taper. If you prefer the larger four-pronged drive, you need to be careful when you're turning the bottom edge of the piece. As you may know, the revolving prongs and turning tools are very incompatible. When using the four-pronged center, I always cut the blank a little longer and leave at least a ½″-long stub or tenon between the prongs and the base of the blank. You should plan the length of your blank accordingly. For those who can afford it, there is also a miniature drive center on the market. The minidrive centers are available in ⅜″ and ½″ diameters, and they are ideal for small spindle projects.

Before mounting the blank, mark the ends for alignment between centers. I always make diagonal lines from the four corners on both ends of the blank. This is an easy method for marking the midpoints of the blank. As discussed in previous chapters, I indent the drive end with an extra four-pronged center before mounting it on the lathe. The job is easy when you have a wooden mallet and a concrete floor. This procedure assures you of good holding power on the drive prongs, and the revolving-center point will easily penetrate the other end of the block.

To turn the blank, I use a 1¼″ roughing-out gouge and a ¼″ or ⅜″ fingernail gouge. True the blank with the roughing-out gouge on its side. You can do the shaping with a fingernail gouge.

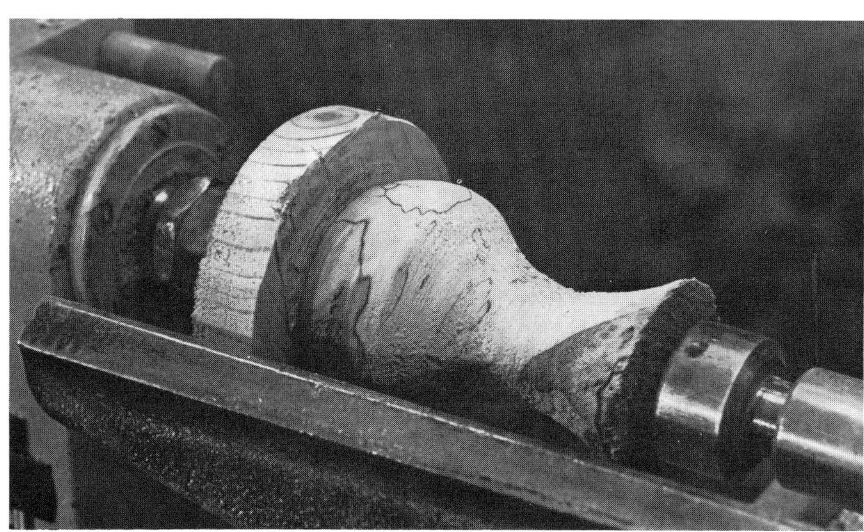

Illus.50. Small spalted weed pot held by double-faced tape

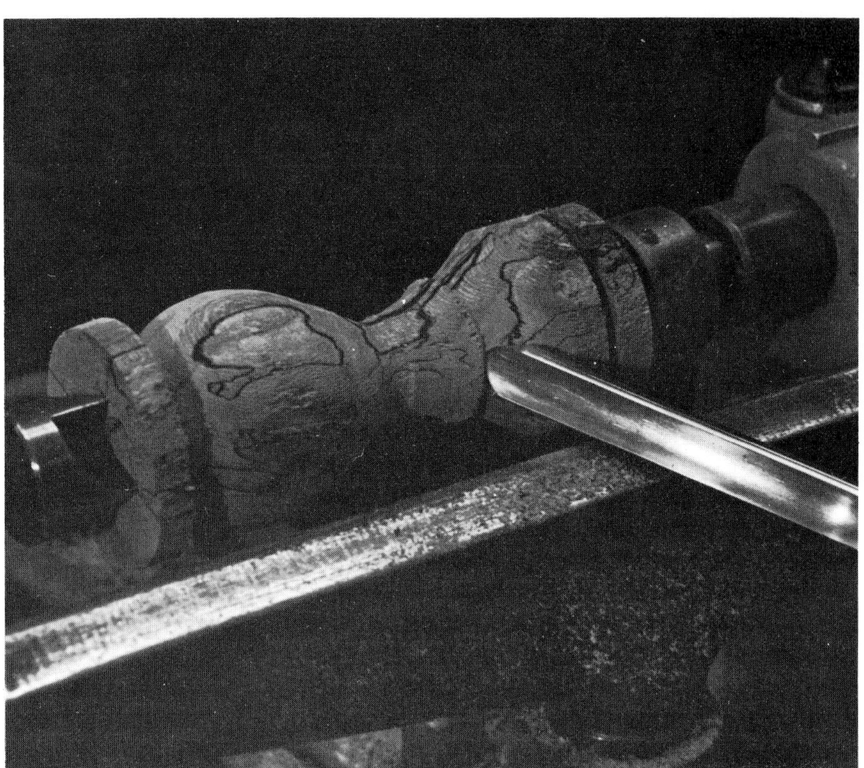

Illus.51. Small spalted weed pot and drive center

When cutting the bottom portion of the blank, be sure to leave a tenon between the drive center and the base. The tenon will serve as a stop for the tool as you roll the bottom surface. A good procedure is to use a parting tool to partially separate and mark the dividing point between the base and the tenon (Illus. 51).

Shape the neck and top lip of the pot as desired. If you want to make some decorative lines around the middle area, use a skew chisel placed upside down on the tool rest and simply move the point into the surface. You can also use a parting tool for this procedure. Hold the parting tool against the wood until it slightly heats up and then burn the cut line around the piece. A few lines circling a small pot can look very attractive.

You may want to cut a narrow bead or two. A small weed pot is an excellent project for doing some very detailed spindle turning. Try out a few different gouges to see what they can do. I like to use a ¼" fingernail gouge on these small projects. You may be very surprised to see the interesting effects you can create. If you happen to ruin a spindle as you're turning beads, simply throw it away and start over again. This is one of the real advantages of turning with scrap material.

You should partially square and clean up the top surface with the revolving center and tenon in place. You can cut off the tenon later on the band saw, or you can separate the pot from the tenon with a parting tool. The parting tool, however, will give you a cleaner surface. When you have completed the turning, go over the surface with a range of abrasives. You may want to shut off the lathe and clean up around any beads you may have turned. If you used good sharp gouges, you may not need to use any abrasives. Remember to ride that bevel.

2. *Regardless of whether you spindle- or faceplate-turned the weed pot*, you will need to drill a hole into it to hold the weed stems. I usually use a ¼" or ⅜" brad-point bit to drill the hole, but the diameter of the drilled hole should be largely determined by the final diameter of the pot's top area. Saw off any remaining tenon(s) and drill the hole using a drill press. Finish-sand both the top and bottom surfaces of the pot. You may want to refer to Chapter 43 for information on finishing products and procedures.

Illus. 1. Small bowls

Illus. 2. Large bowl that had been infested with insects

Illus. 3. Pen holder with paper-clip container, and grimple holders

Illus. 4. Ball-point pens

Illus. 5. Gavel

Illus. 6. Medium-sized bowls

Illus. 7. Toothpick holder

Illus. 8. Small boxes with lids

Illus. 9. Solid apple

Illus. 10. Boxes with inlaid pewter lids

Illus. 11. Bean-pot box with lid

Illus. 12. Large bowls

11
SMALL BOX WITH LID

This design is one of a number of small-box projects that are presented in this book. I'm particularly fond of both designing and turning boxes, especially small ones. As a matter of fact, I'd rather turn a box with a lid than any other lathe project. If you've never turned a box, once you start, you will want to make more. They are great fun to design, challenging to turn, and both functional and decorative. I've found the lathe to be the ideal tool for making boxes of almost any shape and size.

Although I prefer faceplate-turning boxes, you can also easily turn them between centers. There are several spindle-turned boxes in later chapters that I think you will enjoy.

Illus.52. Small box with lid

Almost any kind of wood can be used for boxes. You can use solids or glue up blocks to suit both your taste and pocketbook. In preparing and selecting blocks, you're only limited by

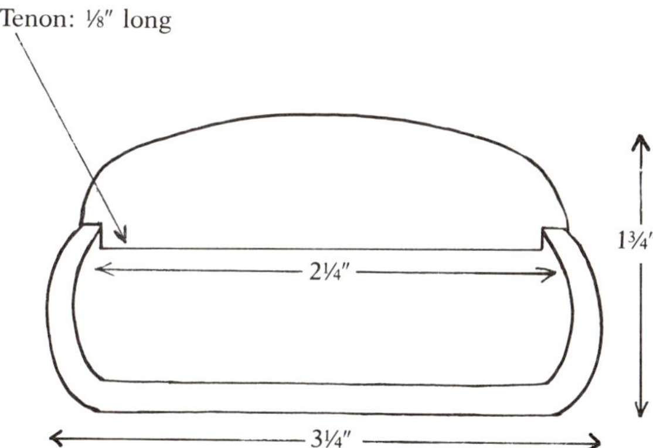

Illus.53. Small box with lid

your imagination. Many of the procedures that are used for preparing blocks for bowls can also be used for boxes. For example, you can use veneers or small solid scraps glued together to form a block for a bowl or box. Methods that are used with one turning project can easily be applied to another.

Since this box project is rather small, you can turn both the base and the lid from 1" (4/4's)-thick stock. Although it's not very deep, the box can accommodate small items of jewelry or other treasures. To assist you in your planning, you may want to look at the base and lid in Illus. 53.

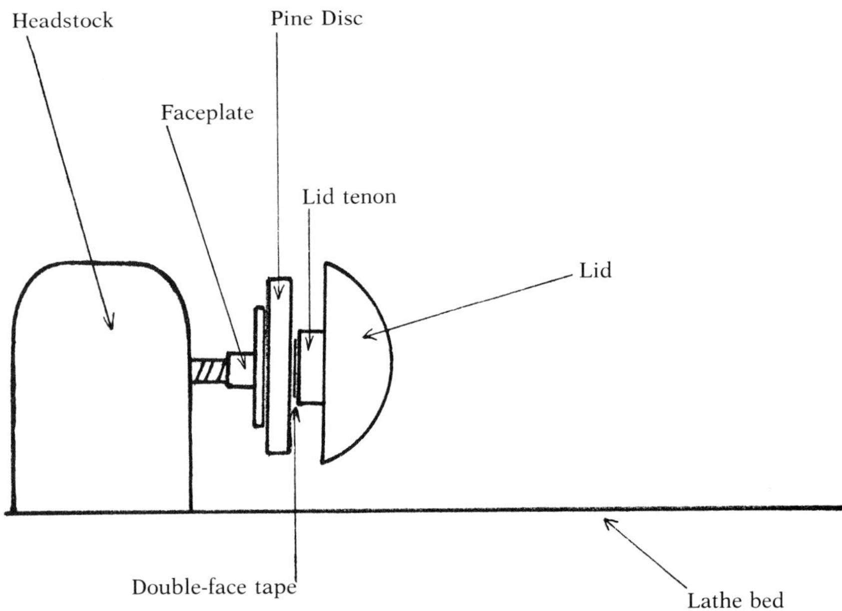

Illus.54. Lid with tenon on faceplate

TASKS:

1. Using a plastic circle template or some other circle-making device, pattern two 1″ (4/4's)-thick pieces of scrap to a diameter of 3¼″. Then cut the blocks on a band saw. While you may want to use glue blocks to secure the base and lid to the faceplate, I prefer using double-face tape. You may want to refer to Chapters 1 and 10 for discussions on the use of the tape.

2. Pattern and cut 3½″-diameter discs from either pine or ¾″-thick plywood. Cut three strips of double-face tape and secure them, side by side, to the disc. Remove the tape covering and center the base block onto the tape. The base of the box needs to be turned first. Press the block into the tape, centered, and then clamp the assembly for a few minutes. Then attach the faceplate to the disc and mount them on the lathe.

3. Rough-turn and shape the base block using a ½″ roundnose scraper. The base edges, both top and bottom, are slightly rolled. Don't worry about cutting into the tape when you are rolling the bottom edge. You may want to refer to Illus. 53 to note the shape of the base.

After you have turned the outer surface of the base, realign the tool rest in front of the block. Using the ½″ roundnose scraper again, turn out the inside of the base. I make the inside wall and bottom slightly rounded. The base opening for the lid tenon should be approximately 2¼″ in diameter. To create more space for storage, I leave a bottom thickness of about 3/16″. The lid has a tenon that takes up some of the storage area in the base; so you want to remove as much wood as possible from the base.

4. When the base has been turned, sand all surfaces using a range of abrasive papers. I often begin sanding with either a 100- or 150-grit abrasive, depending on the type of wood I've used. Finish sanding with a 220-grit and then go over the surfaces with (0000) steel wool.

5. If you have only one faceplate, carefully remove the turned base from the tape. You can use either an old hand plane or wood chisel to force the base from the tape. Be careful that you don't mark the surface with the tool. Slide the blade or chisel between the tape and the base, and then pry them apart. You will quickly discover the strength of the tape bond as you try to remove the base. After you have lifted the base from the tape, remove the used tape from the disc.

Follow the steps detailed in the second task for placement of the lid block onto the taped disc. When you have secured the block to the disc and faceplate, mount the assembly on the lathe.

6. To turn the lid, it's necessary to use a vernier caliper. The lid design includes a ⅛"-long tenon that extends inside the box base. The tenon is what holds the lid in place. The caliper is used to measure the diameter of the base opening and to transfer this measurement to the lid tenon while it's being turned. The caliper is capable of measuring both inside and outside diameters. Once it has been set for the inside diameter of the base, the outside arms are automatically set to the diameter necessary for the tenon.

Before cutting the tenon, I true and reduce the overall diameter of the lid block using the ½" roundnose scraper. It's best to do this initial shaping with the tool rest aligned at the edge and front of the block. After you have properly shaped the lid (Illus. 53), realign the tool rest along the side of the block to cut the tenon.

The ideal tool for cutting the tenon is a parting tool. As I've indicated, I cut the tenon to a length of approximately ⅛". To make the tenon, cut the bottom edge of the block with the parting tool. During this procedure, you will also be cutting away some of the tape, but this seldom presents a problem because the remaining tape is usually sufficient to hold the lid in place (Illus. 54). As you cut the tenon, turn off the lathe and check its diameter with the preset caliper. Also, check the length of the tenon. Continue cutting and checking until you have reached the correct diameter and length. The tenon should fit neatly into the base without being too tight or too loose. To achieve a good fit, the tenon diameter should be a fraction smaller than the actual base opening.

7. After you have turned the lid and tenon, go over the surface with a range of abrasive grits. Slightly round the sharp edge that is on the bottom of the lid proper. Also, slightly round the sharp edge that is left from cutting the tenon.

Using either a hand plane or wood chisel, pry the finished lid from the tape. Test the lid for fit in the box base. Clean up and finish sanding the bottom surfaces of the base and lid. Chapter 43 presents information on various finishing methods and products that you may want to consider.

12
SMALL BOX WITH LID

This chapter presents a small box that is somewhat different from the previous design. As you can see in Illus. 55, the base and lid are made of contrasting woods. The use of contrasting woods in making small boxes is but one of the many designs you can consider. As you read through the box projects in the following chapters, you will discover a whole range of designs that can be used in turning boxes.

I have found that the use of pencil and paper can be extremely helpful in planning box designs. I often make a drawing and include the overall shape of the box and its approximate dimensions, the type of lid I want to use and the way it is secured to the base, as well as any other design considerations that might be helpful. This type of procedure forces me to think through the project and do any problem solving before beginning the actual project. I think you will find that it can be a useful practice before beginning any turning project.

In addition to their range of possible designs, small boxes also have a range of possible functions. You can use them to store small items such as jewelry, or as decorative pieces in themselves. Some box designs that I make for decorative purposes are far too small to be used for anything else. As a matter of fact, the lid tenon almost fills the turned-out base area. You will find instructions for making some of these tiny boxes in a number of later chapters.

The ideal source of wood for small boxes is the scrap pile. You will be surprised at how many different box designs you can make from scrap material. Even wood that is less than ½" thick can be used for turning small boxes. Frequently I use scraps of exotic woods that I've

Illus.55. Small box with lid

saved for this purpose. I'm especially fond of padouk and bocote. It's important to remember that the exotics can be affordable when you're only using small pieces for boxes.

To turn small boxes, I generally use either glue blocks or double-face tape. When turning boxes in quantity, I often use glue blocks to keep the costs down. You can glue up a series of blocks the day before you plan on turning. Using the tape is the easiest method, but it can become somewhat expensive. If you use glue blocks, you can easily saw the finished base and lid from the block using a band saw. So as not to add to the mess of the glue, I don't put paper between the glue disc and the turning block.

I've found that scrapers work best for me when I'm turning small boxes with lids. Generally, a ¼" or ½" roundnose scraper along with a ½" square-nose scraper are all the tools that you need. On occasion I will grind a file or a concrete nail to use as a special tool, but such instances are rare. With some of the spindle-turned boxes in later chapters, you can use a variety of gouges and chisels.

TASKS:

1. Using a circle template or school compass, pattern a block to a diameter of approximately 2½" from stock that is 1¼" (5/4's) thick. Of course, you can vary the thickness of the stock according to what happens to be in your scrap box, and you can increase or decrease the diameter according to your own taste. You need to turn the base first in order to determine the various diameters and thicknesses required for the lid.

Since you are probably only making one box at this time, you might as well use double-face tape if it's available. If it's not, use white glue. Pattern and cut a pine disc to a diameter of at least 3¼". Place two strips of double-face tape in the middle of the pine disc. Since the tape is usually 1" wide, two strips that are 2" long should be adequate for holding the block for this project. After removing the tape's protective covering, center the base block and press it onto the tape. Clamp the assembly for a few minutes, attach the faceplate, and then secure the assembly on the lathe.

2. You can quickly true and shape the block using a ½" roundnose scraper. Illus. 56 shows the shape and details for both the base and lid. After you've shaped the base, align the tool rest so that you can begin work on the top and inside of the block.

I leave a shoulder that is approximately ⅜" wide on the top surface of the base. This not only supports the lid, but also exposes some of the contrasting wood of the base when the lid is in place. You can approximate the shoulder width as you make your initial cut inside the base. To remove wood from inside the base, use a ½" square-nose scraper. You can also use the scraper to square and clean the top surface of the shoulder. While removing wood from the inside, taper the wall outward as you move towards the bottom (Illus. 56). I usually leave a bottom thickness of about ⅛". A small metal ruler is a good device for monitoring the depth of the inside area. Simply compare the inside depth to the actual thickness of the block. Be careful that you don't turn through the bottom—it's been known to happen. Square the bottom surface with the scraper.

3. If the surface needs sanding, use a range of abrasive grits. As a rule, I begin with 150-grit paper, but this depends on the wood that's used. Remember to sand the inside of the base along with the top surface of the shoulder. When you have finished the sanding, carefully pry the turned base from the tape. If you used the glue, saw it off using the band saw.

4. Pattern a 2"-diameter block for the lid from stock that is about 1" (4/4's) thick. As with the base block, you can vary this thickness and diameter according to what's in your scrap box. Cut the block using a band saw. You may be able to use the pine disc you used for the base for turning the lid. Remove any old tape, and replace it with two new strips. Press the lid block onto the tape, clamp, and then secure the assembly on the lathe. You may want to refer to Illus. 56 before turning the lid.

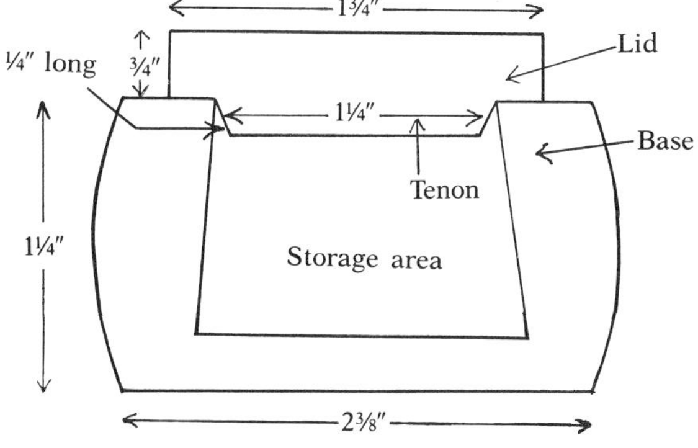

Illus.56. Base and lid

The lid for this project is turned somewhat differently than the base. The top surface of the lid is that portion of the block that is secured by the tape. The tenon of the lid extends outward (Illus. 57). To turn the lid, you may want to use a ½″ square-nose scraper. With this scraper, you can give the top portion of the lid a square wall and also cut and taper the tenon. Before cutting the tenon, measure the inside diameter of the base using a vernier caliper. With the caliper set, use it also to determine the required diameter for the top section of the tenon. To ensure a good fit, make the upper-tenon diameter slightly smaller than the base-opening diameter.

For an added decorative touch, align the tool rest in front of the block. Using either a ¼″ or ½″ roundnose scraper, taper the front surface of the tenon inward. Refer to Illus. 57 to note this area.

5. Sand the lid with a range of abrasives, bringing it to finishing readiness. Carefully pry the lid from the double-face tape. Finish-sand the top surface and lightly roll the top edge. Test the lid for fit in the base. Finish the box as desired or according to some suggestions in Chapter 43.

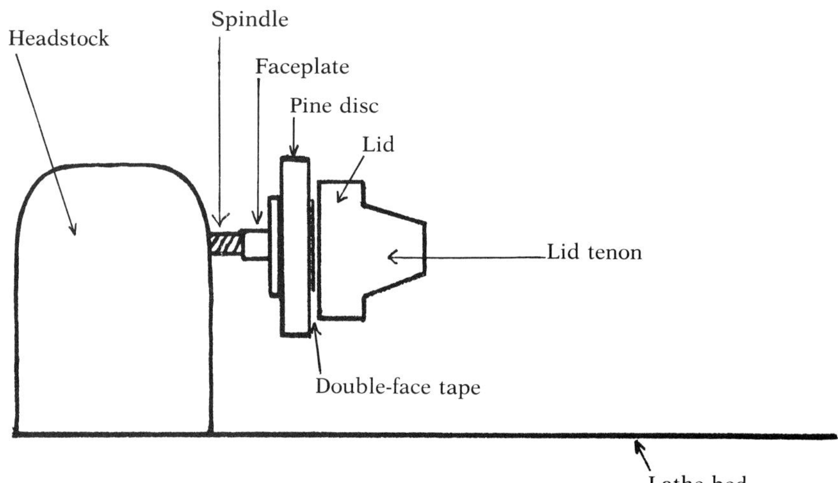

Illus.57. Lid and tenon

13
SMALL BOXES WITH LIDS

As Illus. 58 indicates, you can make this project in different sizes and from different types of wood. The larger box is turned from black walnut and red oak; the other two boxes are turned from bocote, a hardwood that comes from the cordia tree. See how the sapwood of the bocote is highlighted on the lid. This demonstrates another design option you may want to consider as you select wood for the base and lid. When you find anything that is a bit unusual in a piece of wood, try to take advantage of it in your design.

It's also worth noting that the two bocote boxes in Illus. 58 are more decorative than functional because they are so thick. With the lid tenon extending into the base area, there is little room left for storage. The larger box, however, has an adequate area, though small, for placement of tiny items. But a box doesn't need to be used for storage; it can simply function as an interesting, decorative piece in itself.

TASKS:

1. To turn the black-walnut-and-red-oak design, prepare a base block that is about 2½″ in diameter and ¾″ (3/4's) thick (Illus. 59). The lid block should also be about 2½″ in diameter, but a

Illus.58. Small boxes with lids

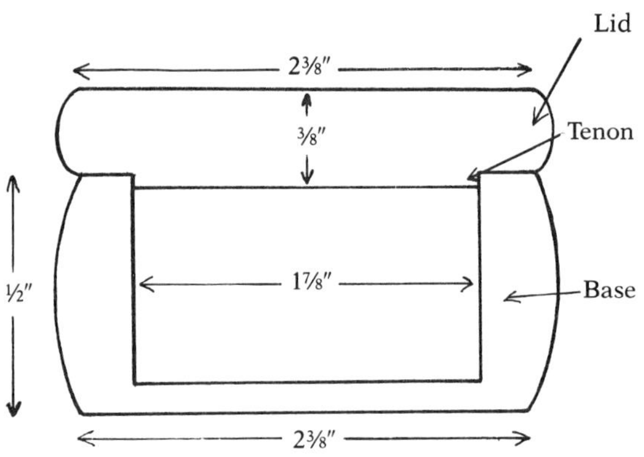

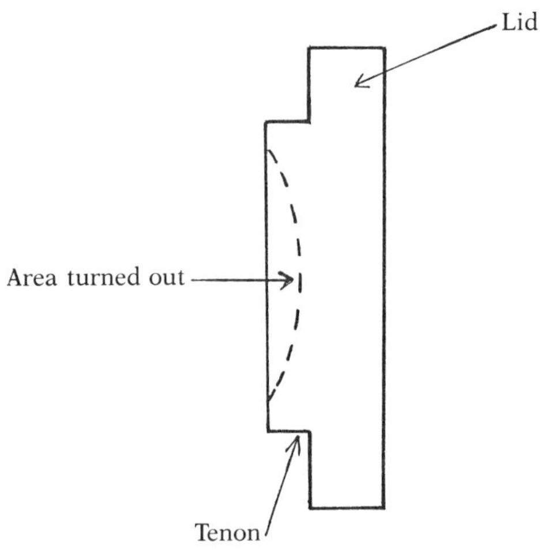

Illus.59. Base and lid

thickness of ⅝″ is more than adequate. Delve into your scrap box and find some pieces that either match or approximate these dimensions.

Both of the bocote boxes have rough block diameters of approximately 2½″. The lids and bases are made from stock that is approximately ½″ (2/4's) thick. However, I often rip a thicker piece when I'm making small boxes from exotic wood (Illus. 60). If desired, you can vary the thickness of the base and lid. Place the surfaces with the saw marks outward so that you can turn or sand them away.

2. If you want to use double-face tape to secure the blocks for turning, refer to Chapter 12 for the discussion on block preparation when using tape. If you don't have any tape, use a small quantity of white glue and secure the blocks directly to 3½″ glue discs.

3. It's best to turn the boxes using ½″ roundnose and ½″ square-nose scrapers. If you tend to have an allergic reaction to some exotic woods, you may want to wear both gloves and a mask when turning bocote. You may also want to wear a long-sleeved shirt if contact allergies seem to be a problem. It's interesting how some of the exotics affect one individual and not another. For example, bocote sawdust often gives me a stuffy nose if I don't wear a mask, but it doesn't seem to bother many other woodworkers. It would be a good idea for you to learn about the potential hazards of various woods and then prepare yourself accordingly.

For the specifics on turning procedures for these small boxes, refer to the discussion in Chapter 12. For finishing procedures, refer to Chapter 43.

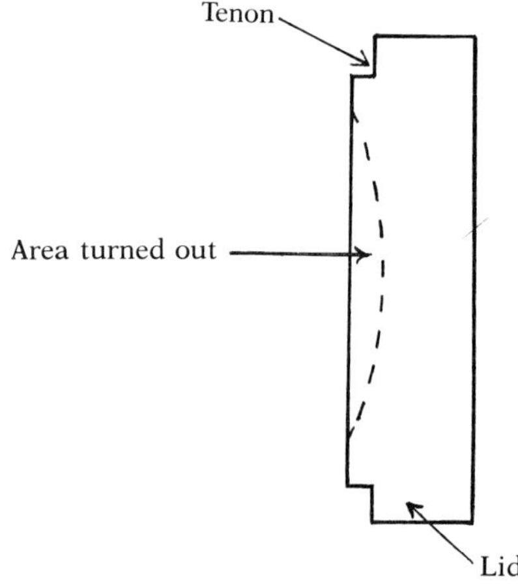

Illus.60. Small box with lid

14
SMALL BOXES WITH LIDS

The boxes in Illus. 61 show how small you can make boxes with lids. While there are a number of designs that I make that are even smaller, once you've learned the procedures for these two, you can then make boxes as small as you wish. By the way, in Illus. 61, the box on the left is turned from coffee wood and the box on the right is turned from bocote.

Although there are miniature turning tools on the market, I've found that standard-size scrapers work quite well for turning small boxes. A ¼″ or ½″ roundnose scraper and a ½″ square-nose scraper are usually more than adequate for the job. On occasion, I've found it necessary to also use a ½″ square chisel when working with small boxes. But you should use it as a scraper instead of a cutting tool. While rather hard on the edge, the chisel can be an effective tool for making boxes. As you turn small boxes with the tools you have available, you will discover for yourself that miniature turning tools really are not necessary. However, you need to keep your tools sharp and use a light touch when you apply them to the wood.

TASKS:

1. For both of the box designs in Illus. 61, you need base and lid blocks that are at least 1½″ in diameter and approximately ¼″ thick. If you use scrap material, you will no doubt have to cut it to the required thickness. It's best to cut the blocks a little thicker than required so that you will be able to sand or turn away the saw marks without significantly reducing the required thickness.

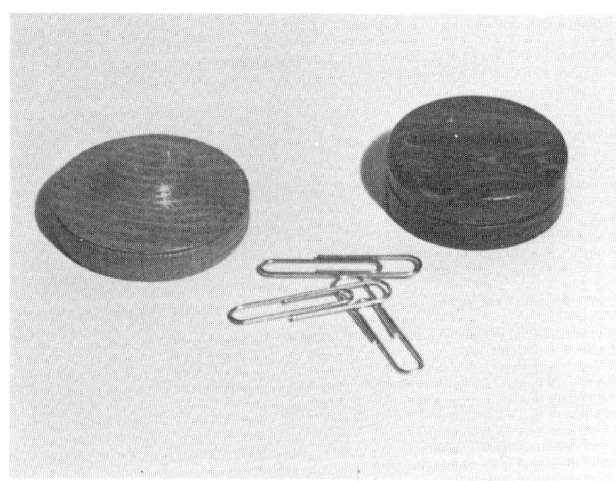

Illus.61. Small boxes with lids

2. Secure one of the base blocks on a 3½″-diameter pine disc with one strip of double-face tape. Center the tape and block on the disc. Usually one strip of tape about 1½″ long is more than adequate. After the block has been clamped for a few minutes, secure a 3″ faceplate to the disc. Now mount the assembly on the lathe.

If the tape is not available, you can simply glue the block to the pine disc. You will lose some thickness in the turned piece because you will need to remove it from the disc using either a parting tool or a band saw. If you target your cut on the glue line, you can minimize any loss in thickness (Illus. 62).

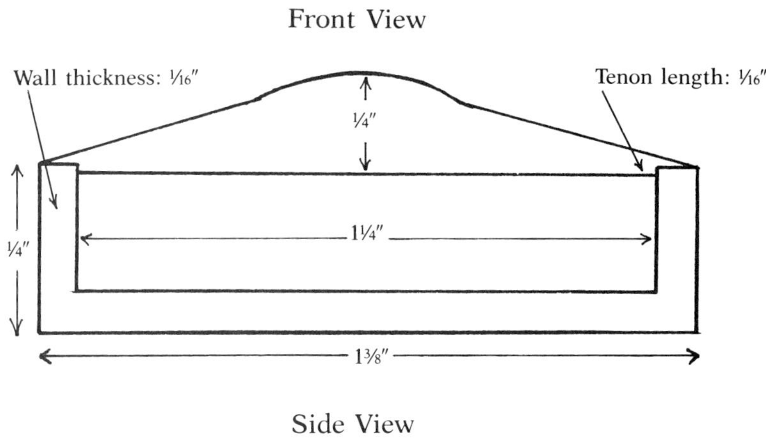

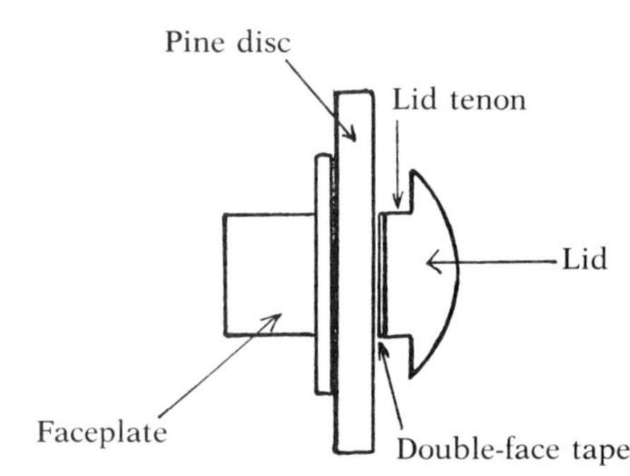

Illus.62. Small base and lid

3. To turn the base, use a ½″ square-nose scraper. First, turn the side surface. True the block and square the outer wall. Remember to use a light touch with the scraper. Then realign the tool rest and turn out the inside of the base with the square-nose scraper. Leave a wall and a bottom thickness that is at least 1/16″. The bottom-inside surface should be flat. You may want to use a ½″ square chisel to make the final cuts on the inside wall and the bottom surface. If the chisel is sharp, it will clean up the surfaces more effectively than the scraper.

4. Once you have turned the base, use fine-grit abrasives on both the inside and outside surfaces. Be careful not to round over the top edge of the wall too much. After you've completed the sanding, remove the assembly from the lathe and carefully pry the turned base from the tape. Take your time with this procedure so that you don't break the finished base. A hand-plane blade or a wood chisel works well for this procedure.

5. Mount the appropriate lid block on the same pine disc. The faceplate can be left in place. You need to use a new strip of tape. As mentioned in previous chapters, the bond will not be strong enough if you reuse the tape. Center the lid block as perfectly as you can on the tape and pine disc. Since you're not working with very much wood, the centering is critical. Then mount the assembly on the lathe.

6. Before you begin turning the lid, set the long arms of a vernier caliper to the outside diameter of the base. The final diameter of the lid must be the same as the final diameter of the base. Thus, use the preset outside measuring arms of the caliper while you turn the lid to diameter.

To true and square the lid's edge, I use a ½″ roundnose scraper. While doing this initial turning, you will also be cutting away some of the tape as you did with the base. But this usually isn't a problem. You should, however, use only a minimum amount of pressure with the tool. While turning the lid block, periodically stop the lathe and check the diameter with the caliper. You want the lid diameter to match the base diameter perfectly. After you have turned the block to diameter, realign the tool rest in front of the block.

As a rule, I use the exposed surface of the block for the lid top and the portion that is attached to the tape for the bottom or tenon area. This allows me to do some shaping on the lid, as indicated in Illus. 62. If you want, you can turn a very small knob on the surface of the lid. Remember, however, to leave enough wood for a tenon that is at least ⅙" long. You can easily shape the top surface using either a ¼" or ½" roundnose scraper. When you've completed the top, realign the tool rest along the side for cutting the tenon.

7. Using the vernier caliper, set the inside prongs to the inside diameter of the base. This will set the outside prongs that you need to use when turning the lid tenon. It's best to cut the tenon with a parting tool.

Since the tenon should be only about 1/16" long, you need to cut into the pine disc and the tape. If you cut directly into the side of the lid block, the parting tool will remove too much wood. As you cut the tenon, stop periodically to monitor its diameter with the preset caliper. As always, you want a fit that is not too loose or too tight. I generally make the tenon diameter a bit smaller than the actual inside-base diameter. If you cut the tenon too long, you can always sand it to the appropriate length.

8. After you have turned the tenon, sand the surface of the lid. Use a fine-grit abrasive so that you don't significantly reduce the diameter or thickness of the lid. When the surface has been sanded, remove the assembly from the lathe. You need to be very careful when you pry the lid from the tape so that you don't inadvertently chip the thin edge of the lid. After checking the fit of the lid in the base, sand the bottom surface of the base and lid tenons. If the lid is too tight, lightly sand around the entire diameter of the tenon using a fine-grit abrasive until you achieve a good fit. For finishing procedures, refer to Chapter 43.

15
SMALL BOX WITH LID

It's the lid in this design that makes this box different from the previous box projects. By employing a thicker lid block, this project demonstrates how you can significantly alter the overall appearance of a box. The box and lid in Illus. 63 have been turned from bocote. I took advantage of a portion of the sapwood in the block to enhance the appearance of the lid knob. This demonstrates how you can maximize anything that is unusual in a piece of wood. The box base in this design has a small storage area.

Illus.63. Small box with lid

TASKS:

1. Pattern and cut two blocks that are 1″ (4/4's) thick and have diameters of about 1½″. If you are using scraps for this project, you can modify these dimensions according to the scraps you have available. However, the base and the lid should be made from blocks that are approximately the same thickness (Illus. 64).

2. Prepare a 3½″ pine or plywood disc, on which you will secure the blocks using double-face tape. Then, center two strips of tape and the base block on the disc. If you don't center the block, you will turn away too much wood. Clamp the assembly for a few minutes. Attach the faceplate to the assembly and then secure it on the lathe.

3. A ½″ square-nose scraper is the best tool to use for truing the base block. The outside wall should be straight, and the square-nose can accomplish this very effectively. On occasion I will use a roundnose scraper and give the outer wall a slightly concave shape. Then I'll give the lid and its knob a similar shape.

4. When you have finished the outer surface, align the tool rest to the front of the block. Using the ½″ square-nose scraper, turn the inside of the base. If you are using one of the exotics, leave the base wall at least ¼″ thick to minimize wasting expensive wood. The bottom thickness should be the same. After you have

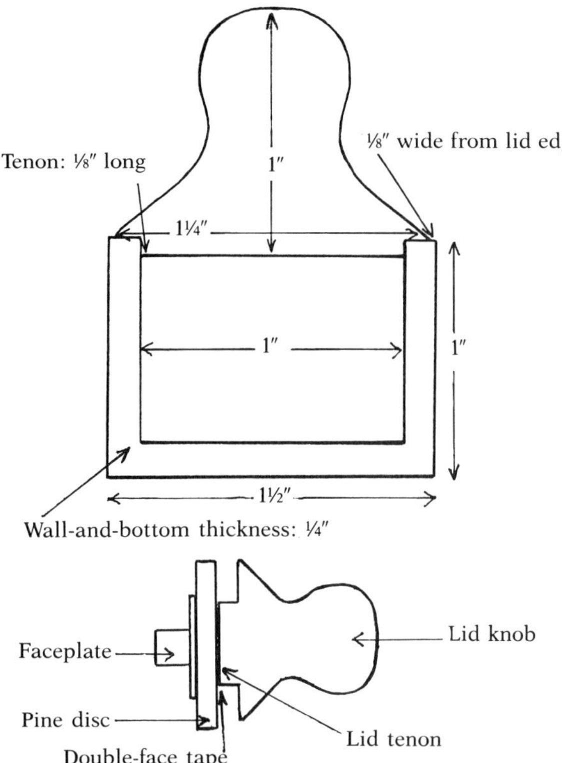

Illus.64. Small base and lid (front and side views)

turned the inside, go over all surfaces with a fine-grit abrasive.

With the assembly removed from the lathe, carefully pry the turned base from the tape. Refer to some of the previous chapters for information on this task. The important thing is to take your time so that you don't damage the turned base.

5. You can mount the lid block on the same disc used for the base. Leave the faceplate attached, but remove the used double-face tape. Cut new strips of tape and place them at the center of the disc and then press the lid block in place. Clamp the assembly for a few minutes. Be sure that the block is centered or it will run off-center and too much wood will be removed. This placement is critical. Now mount the assembly on the lathe.

6. On this particular lid design, it's best to use a ½" roundnose scraper. You may want to review Illus. 63 to note the turned shape of the lid again. To see the various dimensions and how the edge of the lid rests on the top edge of the base, refer to Illus. 64. You should also examine the drawing of the lid mounted on a faceplate in Illus. 64.

Before you true and shape the lid block, set the vernier caliper for measuring the diameter of the bottom edge of the lid. The measurement should be taken from the top edge of the base. You want the bottom edge of the lid to rest approximately ⅛" in from the outer surface. If the base diameter is 1½", the diameter of the bottom edge of the lid should be 1¼". Remember to leave enough length in this area for a tenon that is at least ⅛" long.

True the lid block and then begin some initial shaping using a ½" roundnose scraper. While monitoring the bottom edge with the caliper, turn the midsection of the block concave and roll the top area. The lower end of the curve sweeps to the bottom edge, and the upper end is rolled to form the round knob on top. The bottom edge should taper more radically than the knob area. This edge becomes thin and sharp when you cut the tenon. You may have to realign the tool rest to the front of the block in order to round the knob.

To turn the lid tenon, you must use the caliper again. Set the inside points from the inside wall of the base. This, in turn, will set the outside arms to the appropriate tenon diameter. Cut the lid tenon to a length of ⅛" using a parting tool. To keep from reducing the diameter of the bottom edge of the lid that will rest on the base, cut into the pine disc with the parting tool. Be sure to frequently check the emerging tenon diameter using the preset caliper. This procedure with the parting tool should also leave a sharp edge on the bottom of the lid that will rest on the base.

7. After you have turned the tenon, sand the lid's surface using fine-grit abrasive paper. Be very careful that you don't reduce the diameter of the sharp bottom edge of the lid. This sharp edge should remain intact so that it will neatly taper onto the top surface of the base.

Remove the assembly from the lathe and carefully pry the lid from the surface of the tape. Take your time with this procedure so that you don't chip the thin, tapered edge. Check the fit of the lid tenon inside the base. Using fine-grit abrasives, sand the bottom surfaces of the base and the tenon. Then turn to Chapter 43 for finishing procedures.

16
SPINDLE-TURNED SMALL BOXES

Unlike the previous small boxes with tenon lids that were turned using the faceplate method, this project is made by spindle turning. It also has a tenon lid, but it fits inside a drilled base area rather than one that has been turned. Because of these different procedures, with these boxes you can make rather small diameters. You're limited only by your available drill bits and your ability to turn small tenons. This style of box is one of Andy Matoesian's favorites. The boxes in Illus. 65 are two of his turnings.

I've come to enjoy turning this style of box. I like having the opportunity to use spindle gouges and chisels, and I find the process relaxing. While my boxes are simpler than Andy's, they reflect my taste or, as Andy puts it, my inability to turn small beads. In any event, I think you will find this project an enjoyable change of pace.

Because the wood grain runs lengthwise, or parallel with the lathe bed, in this project, you should use scrap pieces of 1" (4/4's)-thick stock. If you prefer, you can turn the boxes from scrap pieces of dowels. Since dowels are available in a variety of hardwoods, you may prefer using them instead of scraps. A number of later chapters present some very tiny turnings using ½"-diameter dowels and a Jacob's chuck.

While this design is primarily decorative, you can use these boxes for storing tiny items such as sewing needles or pins. Like other small turnings, these boxes seem to be very popular with collectors.

To turn spindle boxes with lids, it's best to use either a standard-size four-pronged drive center or one of the miniature drive centers that

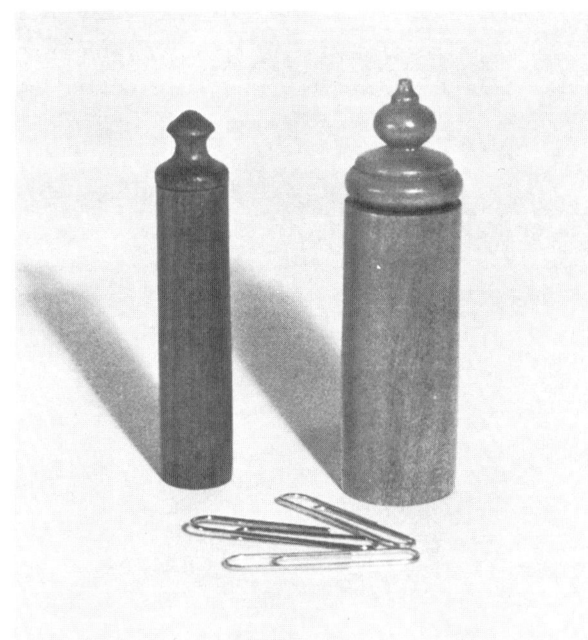

Illus.65. Spindle-turned boxes

are available. A pointed revolving center should be used in the tailstock. If you don't have a ball-bearing center, use the standard dead-center cup with a point that comes with most lathes. I usually extend the point beyond the cut area as far as it will go. This gives better support on the tailstock end of the spindle. It's a good idea to rub some beeswax or paste wax on the point and cut edge. This helps prevent the wood from burning and also facilitates the rotation of the spindle.

In Illus. 66, you can see how the blank is secured between spindles, and you can also see the various parts of the base and lid.

TASKS:

1. Prepare blanks that are at least 3″ long and approximately ¾″ to 1″ square. I suggest preparing more than one blank because I think you will find that these small boxes are great fun to turn and, if you encounter any problems with an initial blank, you will have additional ones ready to use. You should cut the blanks with the grain running lengthwise.

As indicated, it's best to use scrap material for the boxes. You can vary the actual dimensions of the blanks according to the material you have available, but be sure to allow for at least ⅜″ on the ends of the blanks for stubs that will hold the drive and revolving centers. You may want to refer to Illus. 66 to see these support stubs in relation to the lid and the base.

2. For proper centering of the blank on the drive center and the revolving center, draw diagonal lines on both ends of the blank. The lines are drawn from each corner, and where they intersect is the center of the blank.

3. Mount the blank between centers, using the tailstock ram to force the drive and revolving centers into the blank. You may prefer making drive-center indentations on one end of the blank prior to mounting it on the lathe. Using a wooden mallet and the floor as support, center the drive center and then tap it into the blank. This procedure will assure you of a good hold on the drive end. When turning small blanks, this procedure isn't mandatory. It is imperative, however, for you to center the blank properly on the drive and revolving centers. Since you have only a minimum of wood for the box, you can't afford to waste any of it by having the spindle placed off-center.

4. Using a roughing-out gouge or a roundnose scraper, bring the blank to roundness. If you are using a roughing-out gouge, lay it on its side. I find this to be a more efficient way to round a blank. Be certain to ride the bevel. If you are using a standard drive center, be careful that you don't hit it with a tool. It's not necessary for the end stubs to be rounded.

5. Mark the inside edge of the stubs using either the point of a skew chisel or a parting tool. As indicated, the stubs should be at least ⅜″ long. This length should provide enough wood for

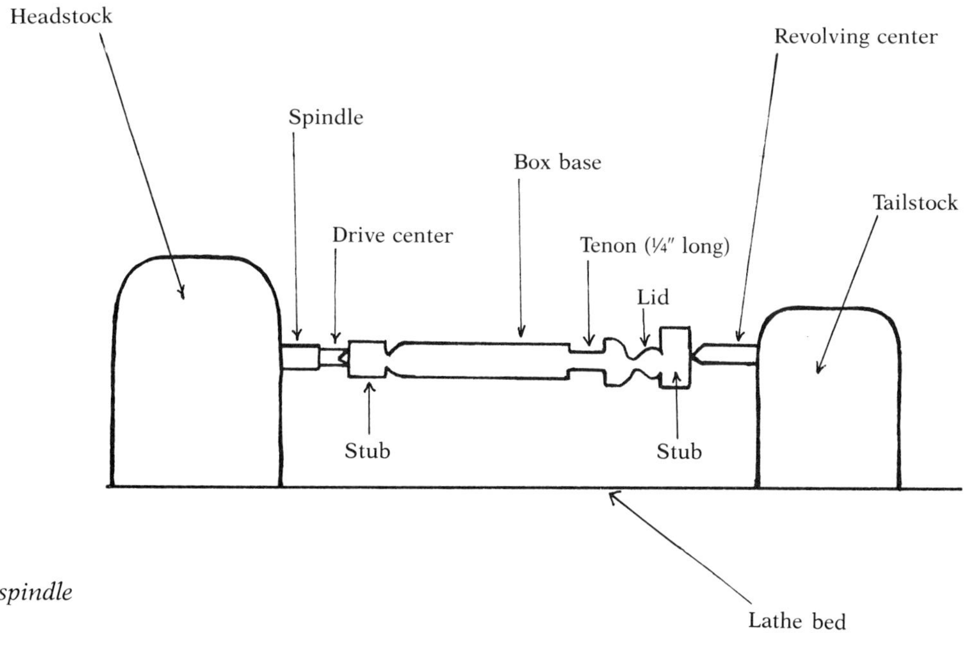

Illus.66. Box-and-lid spindle between centers

Illus.67. Turning lid of small spindle box

both the drive and revolving centers, and it shouldn't reduce the overall length of the box too much.

6. On a small box that will have a finished diameter of ¾", I generally cut the lid tenon to a length of approximately ¼". The tenon should be slightly less than ½" in diameter. Determine the approximate length of the base you want in the final box. Be sure to allow enough wood for the lid. The tenon needs to be cut between the planned base length and the lid (Illus. 66).

Since you want a good fit of the lid tenon in the base, you will need to use a vernier caliper to monitor the tenon's diameter. You will be drilling out the center area of the base using a ½" brad-point bit. Thus, you must turn the tenon to a diameter of slightly less than ½".

To cut the lid tenon, use a parting tool. Move the parting tool from side to side, frequently stopping the lathe to check the diameter of the tenon. You should turn the tenon slightly smaller than the planned base hole. If the tenon diameter is too large, it will not fit into the drilled base area. If it is too small, the tenon and lid will flop around when placed on the base. As indicated, on a ¾" diameter box, make the tenon approximately ¼" long. If you plan on drilling a ½"-diameter hole in the base, turn the tenon slightly smaller than ½" in diameter.

7. After you have made the tenon, it's time to turn the lid. You may want to refer to Illus. 65 to see some possible lid designs. To turn the small beads or coves on the lids, Andy and I both like to use a ¼" fingernail gouge that is ground slightly long. However, you may prefer to use a ¼" roundnose scraper to turn the decorative lid (Illus. 67).

As you turn the lid surface, do not separate the lid from the stub. You can do this procedure after you have sanded the surfaces of the base and the lid. If your gouges or scrapers are sharp, you may not need to do too much sanding. In any event, use a fine-grit abrasive for any necessary sanding.

8. There are a number of ways to separate the lid tenon from the base and the two stubs from the lid and base. You can remove the spindle from the lathe and then cut the three areas using a band saw. In other words, you would be sawing the stub from the base, separating the lid tenon from the base, and then sawing the tailstock stub from the top of the lid.

Using a gouge or skew, another option is to

reduce the connecting diameter between the top of the lid and the tailstock stub to about 1⁄16″ in diameter. With this diameter, you can break the stub off the top surface of the lid when you remove the spindle from the lathe. Separate the lid tenon from the base using a parting tool. Let the side of the parting tool rub against the top edge of the base as you make the cut. This will assure you of a good flat surface, against which the lid will fit snug and flush (Illus. 68). The lid and base sections will fall from between the centers when you complete the separating cut. Try to catch both pieces because their surfaces could get damaged if they dropped onto the lathe bed. Now break the stub off the top of the lid. Using a band saw, cut the other stub from the bottom of the base. Be sure you have a flat, square surface on the bottom of the base so that the box will stand erect.

Regardless of the option you chose, you should now lightly sand the top area of the lid and the bottom edge of the base with a fine-grit abrasive.

9. The remaining task is to drill the storage hole in the base. The hole should penetrate the exact center of the base; so you need to find the center and mark it for drilling.

There are a number of devices you can use that can help you find the center of a round object. You can also mount the base in a commercial chuck and then mark the center with a pencil as the assembly revolves on the lathe. However, Andy and I prefer to use the "eyeball" method for finding the center. Simply approximate the center of the top end of the base and mark it with a pencil. Use this pencil mark as the entry point for the bit.

As I've indicated, you should use a brad-point bit of the appropriate diameter to drill the storage area in the base. To make sure you get a straight, accurately drilled hole, use a drill press for this procedure. Set the drill press so that you will have a bottom thickness of approximately 1⁄8″. Remember to anticipate the added depth made by the point of the bit. To prevent the base from rotating during the drilling process, I generally wear a leather glove to get a good grip. If you prefer to use a vise to hold the box base in place for drilling, be careful so that you don't mark the surface of the base with the vise. Drill the hole to the appropriate depth.

10. With the storage area drilled into the base, test the lid tenon inside the hole for fit. You ought to have a good snug fit. If the tenon is too tight, lightly sand around the surface of the tenon with a piece of fine-grit abrasive paper. Continue testing and sanding until the lid and its tenon rest properly on the base. Then refer to Chapter 43 for finishing procedures.

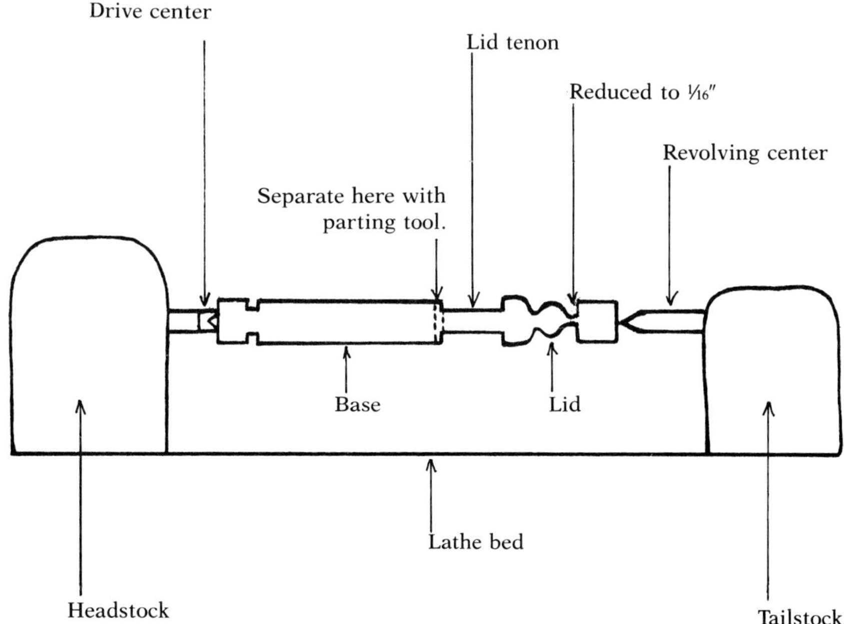

Illus.68. Separating lid tenon from base

17
SPINDLE-TURNED BOXES

This project is a variation of the project presented in the previous chapter. Like the boxes in Illus. 65, the ones in Illus. 69 were turned by Andy Matoesian. The lid for this design employs a tenon that penetrates into the base storage area. You may want to refer to Illus. 66 for specifics on securing and turning this type of design between centers.

While this box design can be turned from scraps, you may want to consider using regular stock if you use this design for larger boxes. The design lends itself to turning, with the grain, boxes of any diameter. The boxes in Illus. 69, for example, range from 1½″ to 2¾″ in diameter. If you don't have wood this thick, laminating wood can be a good alternative. By the way, the

Illus.69. Small spindle-turned boxes

boxes in Illus. 69 range from 2½″ to 3½″ in length.

To achieve a particular stock thickness, you can laminate 1″ (4/4's)-thick stock together. By using different woods, you can achieve a very interesting and unusual effect. Another option is using a few strips of veneer. Glue a strip of contrasting veneer between one or two pieces of 1″ (4/4's) stock. The veneer can be especially attractive in a lid when you add a few beads and coves. Since the design requires that you turn both the lid and the base from the same piece of wood, laminating can have a very striking effect.

In addition to turning the project from laminates or scrap material, you can also use small logs from your woodpile. Find some logs with a 3″ to 4″ diameter and a length that is suitable for your box design and then turn them with the grain running parallel to the lathe bed. If sufficiently dry, small logs can be an excellent and endless source of wood for the spindle turner.

The various tasks for this project are essentially the same as those presented in Chapter 16. Because the internal diameter of the base is larger in this design, you will need to use multispur bits. For example, 1″-, 1¼″-, and 1½″-diameter multispur bits can be very effective for drilling out the storage areas in these types of boxes. To avoid having to buy these rather expensive bits, you can also easily turn the internal base area using a ½″ square-nose scraper. In order to tool out the internal storage area, you will need to mount the blank on a glue or tape block. After you have turned the lid and its tenon, remove the lid from the base using either a parting tool or hacksaw. You can also remove the entire assembly and cut the lid and its tenon from the base using a band saw. Remount the base and turn the internal area to match the tenon diameter. As with the boxes in the previous chapter, you should use a vernier caliper to determine the appropriate diameter.

Before you begin this project, you may want to do some initial planning and design work. This could be especially helpful in relation to how you make the lid.

Now follow the project tasks presented in Chapter 16.

18
SMALL BOXES WITH LIDS

This project begins a series of small-box designs that have a shoulder cut into the base to hold and support the lid. You don't need to use lid tenons with these designs. And this type of design requires you to do faceplate turning as opposed to spindle turning. As you begin turning boxes with lids, you will discover any number of ways the lid can fit on or into the base. You will also discover some very interesting and clever ways for the lid to be secured on the base if you do some pencil-and-paper design work. This project demonstrates one way that is relatively easy and certainly effective, and it lends itself to both small and large boxes with lids. A number of later chapters present large-box designs that employ this type of lid arrangement. The boxes presented in this chapter and shown in Illus. 70 were turned by hobbyist turner Chris Weggemann.

TASKS:

1. Pattern and cut two blocks that are ¾″ thick and at least 3″ in diameter. However, you may want to vary these dimensions based on available stock. You can easily turn these boxes from

Illus.70. Small boxes with lids

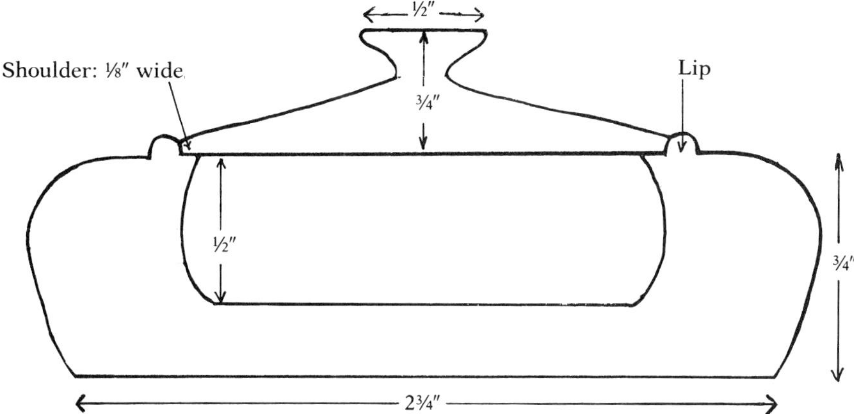

Illus.71. Base and lid

scrap material or, if necessary, glue together thinner stock to achieve the approximate thickness.

You should use double-face tape and pine discs to turn both the base and lid for the project. For procedures on using double-face tape for faceplate turning, you may want to refer back to Chapter 14. Incidentally, with this type of design, you need to turn the base of the box first.

2. To assist you in turning this project, see Illus. 71, which depicts both the shape and approximate dimensions of the base and lid. As you can see, there is a small knob on the lid. Illus. 72 shows another lid design that you can use. This alternate design has a square edge area sanded into the lid, which permits the user to remove the lid by prying it off with a fingernail.

For turning the base, Chris uses a ½″ or ¼″ roundnose scraper. For making the lid shoulder and the storage area, he uses a ½″ square-nose scraper. He also uses a ½″ square-nose chisel as a scraper to clean up the shoulder and the inside surfaces of the base.

True and shape the base block using the appropriate tools. After you have shaped the side wall, align the tool rest in front of the block. As you can see in Illus. 71, a lip goes around the top surface of the base. You should cut the lid shoulder on the inside of the lip. The shoulder should be recessed about ⅛″, and it should also be at least ⅛″ wide. The lid then fits into this recessed area and rests on the shoulder. A ½″ roundnose scraper is an ideal tool for making the lip, the inside recess, and the lid shoulder. Now cut the lip and the shoulder.

After you have made the shoulder and the lip, remove wood from the inside of the base and at

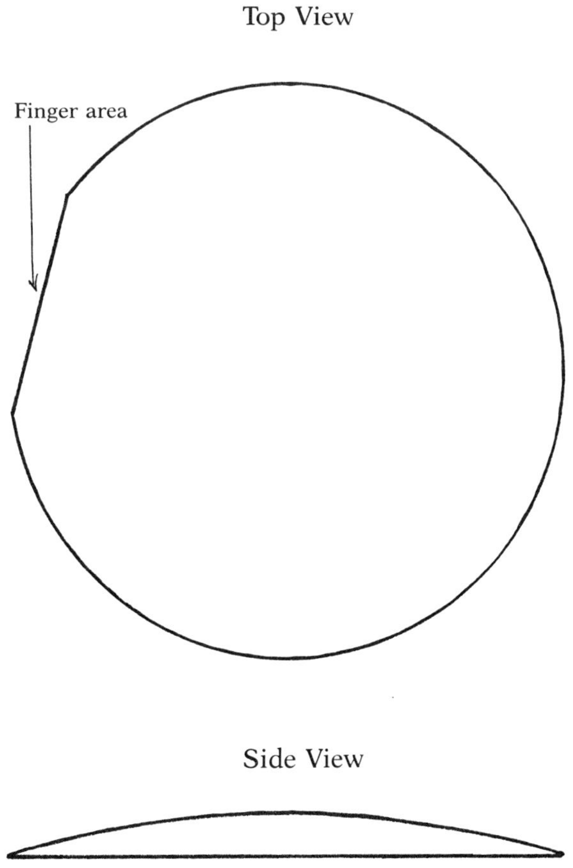

Top View

Finger area

Side View

Illus.72. Lid design

the inside edge of the shoulder using a ½" square-nose scraper. Remember to leave the shoulder at least ⅛" wide. You should cut the inside wall straight down to the inside-bottom surface. Leave a bottom thickness of at least ⅛". Be certain you periodically check the inside depth in relation to the actual thickness of the block. A small ruler is a good device for this procedure. You don't want to turn through the bottom surface. Make the bottom surface flat.

3. When you've turned the base, go over the shoulder and inside surface with a ½" square-nose chisel used as a scraper. This procedure cleans up the various surfaces and also sharpens the corner between the lip and the shoulder. Lightly sand with a fine-grit abrasive paper, depending upon the overall quality of the surfaces. Be careful when you are sanding the lip that surrounds the top surface. Sand off the sharp edge on the top of the lip, but don't significantly reduce the lip's height. After you have sanded the surfaces, remove the assembly from the lathe and carefully pry the turned base from the tape.

4. You can center the lid block on the same pine block you just used for the base. Remove the used double-face tape, but leave the faceplate attached. Double-face tape tends to be effective for only one use. Apply new strips of tape to the pine disc, center the lid block, and briefly clamp. Then mount the assembly on the lathe.

5. It's best to turn the lid using ¼" and ½" roundnose scrapers. You can easily turn the knob on the lid (if you prefer this design), using the ¼" roundnose scraper. You may want to refer to Illus. 71 to note the shape of the lid. You need to use a vernier caliper for turning the lid.

Set the caliper prongs to measure the inside-lid recess in the base. You want the measurement to be from the inside surface of the lip. This procedure will automatically set the outside arms of the caliper. Using the most appropriate scraper, true the lid block and reduce it to the required diameter. Check the diameter often with the preset calipers. Don't worry about cutting into the tape on the pine block since the remaining tape will usually be adequate for holding the block in place. If you're concerned about the block coming off, use a revolving center and the tailstock as a support. You may want to place a small piece of dowel between the point of the revolving center and the top surface of the block to prevent the point from penetrating the surface of the block.

Beginning at the outer edge, taper the top surface of the lid with a ½" roundnose scraper. The knob should extend at least ⅜" above the lid's surface. The final thickness of the lid's edge should be the same as the depth of the lid recess in the base. The top surface of the lid tapers upwards towards the knob. Using the ⅛" roundnose scraper, make the knob in the lid's surface. Chris usually makes the top portion of the knob at least ½" in diameter. Be careful that the diameter at the base of the knob is not too small or else the knob will break off during the turning process. You can easily finish the top surface of the knob by aligning the tool rest in front of the block. Use the ½" roundnose scraper and a very light touch to finish the top surface of the knob.

6. When you have completed the turning, sand the entire surface with a range of fine-grit abrasive papers. Sand the neck of the knob very carefully because too much pressure could easily break it off. Then you can go over the surface with (0000) steel wool.

Remove the assembly from the lathe and then very carefully pry the lid from the tape using either an old plane blade or a sharp chisel. Be especially careful that you don't chip out portions of the lid's thin edge. After you have removed the lid from the tape, test the lid for fit in the base recess. If the lid is too tight, lightly sand the edge of the lid until it fits. Use a very fine abrasive for this procedure. Sand the bottom surfaces of the base and the lid, and then turn to Chapter 43 for finishing procedures.

19
SMALL BOXES WITH LIDS

This project presents another small-box design with a recessed lid shoulder. Although the design is small enough to interest people who collect small boxes, it's also large enough to provide some space for storage. With this project, there is sufficient wood for the lid so that you can also turn it round or with a tiny knob. From left to right, the boxes in Illus. 73 are turned from scrap pieces of padouk, bocote, and ebony.

As with many of these smaller boxes, it's best to turn this design using a 3″ faceplate, a 3½″ pine disc, double-face tape, and scrapers. If you are hesitant to use the tape, you can glue the turning blocks directly to the pine disc. For specifics on turning this project, refer back to the instructions in Chapter 18. For assistance in planning and turning this type of design, see the diagram of the box in Illus. 74.

Illus.73. Small boxes with lids

The project requires two blocks cut to a diameter of 1¾". The stock should be approximately 1" (4/4's) thick. However, you may want to vary these dimensions according to your available material. With this design, you can use any kind of wood you want. Although the boxes in Illus. 73 are made from imported woods, I usually turn this design from walnut, wild cherry, or oak. I'm sure you have all kinds of wood in your scrap box that can be used for boxes.

I want to emphasize again that the dimensions for this project are only suggestions to assist you in making an initial box. I think you will find, as I have, that it's great fun to see how small you can make a box. The lid recess and shoulder in this design lend themselves to very tiny boxes. While I generally use standard-size scraping tools, on occasion I've had to make a special tool from a concrete nail or a small file. I've found that scraping is the most effective method for turning these small boxes.

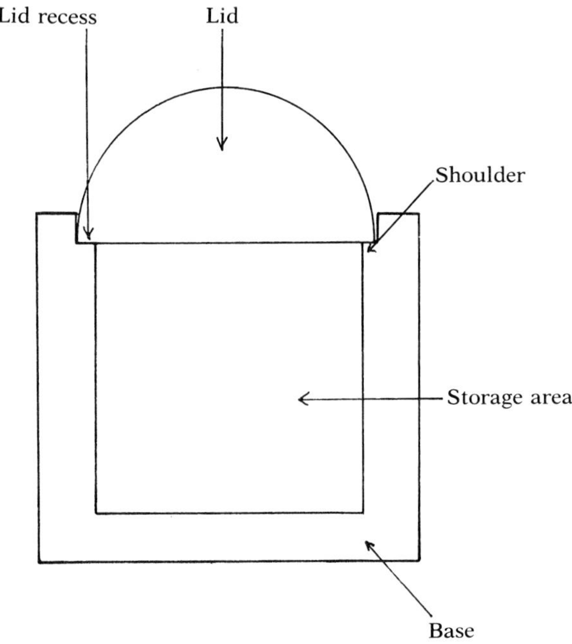

Illus.74. Base and lid

20
SMALL BOX WITH LID

Like the boxes in the previous two chapters, this box also has a recessed lid shoulder, but the box is considerably larger than the other two designs. It's actually an ideal size for a box because it provides some storage area but yet isn't so large as to overwhelm any area where you might place it. This box is quite decorative and it's clearly more functional than some of the other small boxes.

I turned the box in Illus. 75 from a piece of severely checked bocote, which I had tossed in my scrap box. I turned the lid round to showcase one of the larger checks. The natural aberrations in wood can make a piece very interesting.

Although the box in Illus. 75 has a rolled lid, you may prefer to turn it with an appropriate-size knob. One drawback with rolled lids is that they are not as easy to remove as lids with knobs. To facilitate removing the lid from the base, you can square a small section of the lid edge so that the user can open or remove the lid with a fingernail. You can see a drawing of this kind of lid in Illus. 72. Another option for quick removal of a rolled lid is to keep the lid recess in the base rather shallow. By applying pressure to one side of the lid, the lid will then slide out of the recessed area. You may want to refer back to Illus. 74 to note how the lid recess and shoulder should be made for this design.

For this project you need to prepare two blocks that are at least 3½″ in diameter. For the base block you should use stock that is about 1¼″ (5/4's) thick, and for the lid block you should use 1″ (4/4's)-thick stock. If you don't have scraps of these dimensions, you may have to use some new stock for this project. It may be helpful to see Illus. 76 on page 90, which depicts the base and lid along with some approximate dimensions.

Illus.75. Box with round lid

I turn the base and lid using double-face tape mounted on a 3½″ pine disc and a 3″ faceplate. For best results, I use a ½″ roundnose scraper as well as a ½″ square-nose scraper. Now refer back to Chapter 18 for turning instructions.

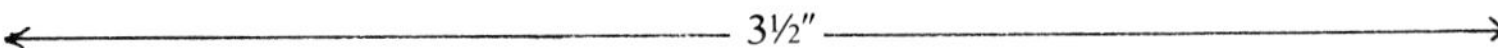

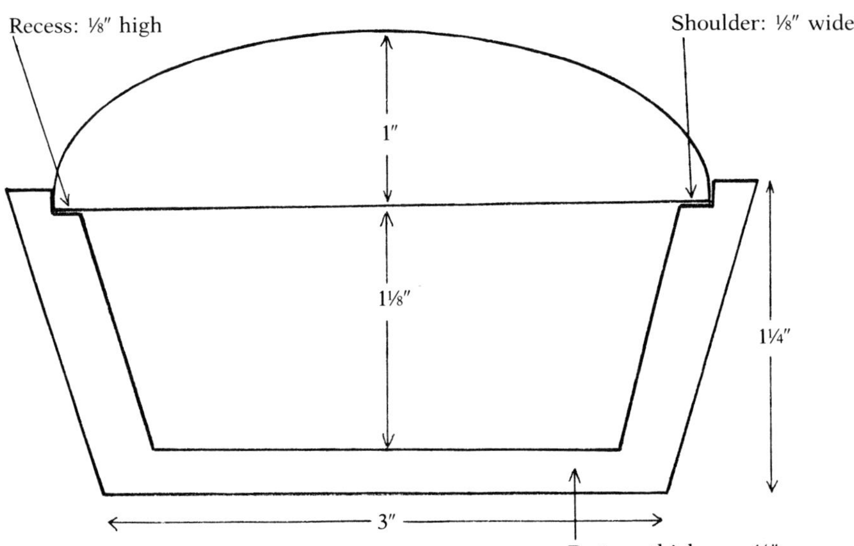

Illus.76. Base and lid

21
LARGE BOX WITH LID

This chapter presents a design and various procedures for turning a rather sizable box with a lid. Because of its size, the box provides a substantial storage area. However, the design of the box makes it a decorative piece as well. This project requires standard stock for both the base and lid, but you can easily prepare the lid knob from a piece of scrap. The box and lid in Illus. 77 have been turned from black walnut.

For some assistance in preparing for and turning this project, it would be a good idea for you to review the various discussions in Chapter 1. You should turn the project using a glue disc and a 3″ faceplate. Although you can use double-face tape, I prefer using the glue-block method with a block this size. Illus. 78 shows the project design along with some approximate dimensions. So that the lid will remain securely in place on its shoulder, I make a 3/16″-long tenon on the bottom of the lid (Illus. 79).

Illus.77. Large box with lid

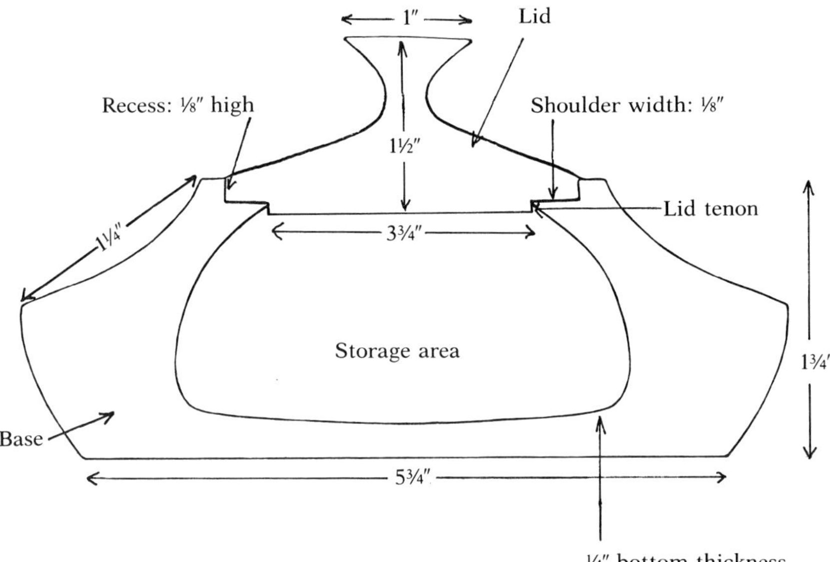

Illus.78. Base and lid

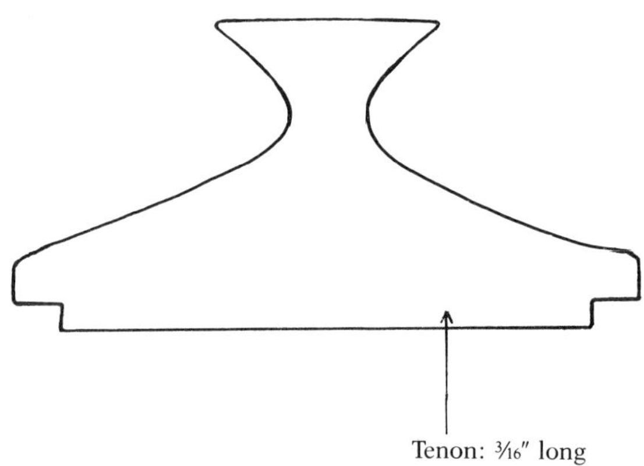

Illus.79. Lid with tenon

TASKS:

1. Prepare a base block that is 2″ (8/4's) thick and at least 6″ in diameter. The lid block should be prepared from 1″ (4/4's)-thick stock and have a diameter of at least 4″. It's important for you to remember the lid-block diameter as you turn the base, the storage area, and the lid recess, because you don't want the opening to exceed the diameter of the prepared lid block.

Cut a lid-knob block that is at least 1″ (4/4's) thick and has a diameter of approximately 1¼″. If you prefer a longer knob, use wider stock. Also, using a wood for the knob that is different from the wood used for the base and lid can be a nice touch. Center the knob; then glue and clamp it to the top surface of the lid (Illus. 3). You may want to review the discussion of this procedure in Chapter 1.

2. Center a 3½″ pine or plywood disc on the bottom of the base block, glue the disc and the base block together, and then clamp. So that the lid block will be ready for tuning when the base block is done, glue a disc to its base. Be sure the glue used to secure the knob to the lid block's surface is sufficiently dry before unclamping them. You may want to review the discussion on using glue blocks in Chapter 1.

After the glue is dry, secure a 3″ faceplate to the disc on the base block and mount the assembly on the lathe.

3. I use a 1″ roundnose scraper to true the block and do some initial shaping. This is a good time to refer back to Illus. 77 to see how the finished project looks. You should also consult Illus. 78 to get some help with the shaping process and to note the various dimensional requirements of the base.

Using either a 1″ or ½″ roundnose scraper that's been freshly ground, do the final shaping on the side-and-bottom edge. I tend to roll the bottom edge slightly. Given the size of the block, you have ample room for using the scraper on this edge.

Align the tool rest to the front of the block and then shape the top surface of the block. In Illus. 78, you can see how this surface has been turned slightly concave. This area should be at least 1¼″ wide to create the desired appearance in the base. Also, this width will ensure that the lid recess and storage area are not too large for the prepared lid block. Make a few light cuts with the scraper to clean up the surface of this area.

4. With the lathe running, mark the middle of the block by simply placing the edge of a scraper or the point of a skew chisel against the approximate center of the block; then the tool and lathe will do the rest. Now shut off the lathe. Using a ruler and pencil, lay out the diameter of the lid-and-storage opening. Make dark pencil marks on the surface so that you can see them while the block is rotating. If you prepared a 4″-diameter lid block, this marked diameter should not exceed 3¾″. This will allow the lid block to be reduced during the turning process and still fit into the base.

5. Using a ½″ or ¾″ square-nose scraper, begin removing wood from the storage area of the base. You should begin removing it well inside the marked diameter. The pencil marks should be highly visible as a guide during this procedure. After you have removed some of the wood from the storage area, cut the lid recess at the marked diameter line using a ½″ square-nose scraper. The recess should be approximately ⅛″ deep. Next, cut the lid shoulder by simply forcing the scraper into the wood of the storage area approximately ⅛″ from the recess.

In order to round the inside wall and bottom, I use a ½″ roundnose scraper. You should also remove the remaining wood with this tool. You can leave a bottom thickness of approximately ¼″ or less, if desired. Be sure you don't cut off the lid shoulder while removing wood from the storage area.

6. After you've turned the base block, sand all surfaces (if they need sanding) using a range of abrasive grits. Be very careful around the lid recess and shoulder so that you don't reduce their size. You may want to go over them very lightly with a 220-grit abrasive. You should also sand the inside surface of the base. As a final preparation of the surfaces, I use (0000) steel wool.

7. There are a number of ways to remove the glue disc from the bottom surface of the turned base. One option is to turn it off using a parting tool. If you use this procedure, don't cut all the way through the disc. Leave a small attaching tenon that is at least ⅛″ in diameter. You can then break it off from the base. On occasion, I will cut all the way through the glue disc with the parting tool and then catch the base as it drops towards the bed. Unfortunately sometimes I miss the base and it falls on the lathe bed or the floor, which marks its outer surface. It's a great trick, however, when it works. Incidentally, you should keep the side of the parting tool tight against the bottom surface of the base. This way, you will cut away most of the glue that is on the bottom surface.

Another option for removing the glue disc from the base involves using the band saw. To prevent the band-saw blade from pulling the assembly from your hands, you should support the glue disc with a block of wood. I also have the faceplate attached so that I will have more to hang onto during the sawing process. The saw blade should be cutting along the bottom surface so that you will also remove any mess from the glue. Take your time with this procedure so that you don't damage the turned base.

After you remove the glue disc, clean up the bottom surface on a belt sander. Then bring the bottom surface to finishing readiness by using a small pad sander or by sanding it by hand. Use the same range of abrasive grits you used on the other surfaces of the base. Then rub the surface with (0000) steel wool.

8. Secure a faceplate to the lid block and then mount the assembly on the lathe. Before you begin to turn the lid, determine its final diameter by measuring the base's lid-recess area with a vernier caliper. Use the preset caliper frequently to check the lid diameter while you are turning the lid. With this type of base-and-lid design, the lid diameter is critical.

To prevent the lid knob from breaking off during the turning process, you may want to bring the tailstock with a revolving center forward. This will sufficiently support the knob so that it won't break off.

9. True the lid block using a 1″ or ½″ roundnose scraper. As you true the block, begin rough-shaping the top surface. You may want to refer back to Illus. 79 to see how the lid is shaped. It's especially important to note how the edge is tapered to fit neatly into the recess but also to be flush with the top surface of the base. Monitor the diameter with the preset caliper.

When you have trued and turned the lid block to the appropriate diameter, you need to cut the 3⁄16″-long lid tenon. I've found that the ½″ square-nose scraper is the ideal tool for making the tenon. Before you make the cut, set the vernier caliper to the inside diameter of the storage area. The prongs should be set against the inside edge of the lid shoulder. As you know, the tenon must fit neatly inside this area. While monitoring its diameter with the preset caliper and its length with a ruler, cut the tenon with the square-nose scraper. You should make the tenon a fraction longer than needed in anticipation of some wood removal from its surface because of the glue and the disc.

10. Before you do the final shaping of the lid's top surface, it's best to turn the knob. Align the tool rest as near the knob as possible so that you will have good tool support. Too much chatter from a poorly supported tool can cause the knob to break off, even if it's supported by the tailstock. Also, the tool chatter will cause the knob to have a rough, untidy-looking surface.

I recommend using a ½″ roundnose scraper to make the knob. Make the initial cuts along the glue line, where the knob and the lid block come together. You need to work on this area so that the glue line becomes almost invisible. Cutting the rounded base of the knob is very similar to cutting any cove: You should move from the outer edges to the middle. You may want to refer back to Illus. 77 to note the appearance of the finished knob. Now finish turning the side of the knob by cutting very lightly with the scraper. You can turn the top surface of the knob after you've removed the tailstock.

11. Realign the tool rest and finish turning the top surface of the lid. Remember that the edge should taper and it should not extend over the recess area. You want the lid to flow evenly into the top surface of the base. I tend to give the lid a slightly concave shape between its edge and the bottom of the knob. You can easily create this shape with a ½″ roundnose scraper.

12. Remove the tailstock and align the tool rest near the top surface of the knob. Using a ½″ roundnose scraper, make the knob's top surface slightly concave. Don't use too much pressure with the scraper, or you could break off the knob. And be sure you turn away the indentation made by the point of the revolving center.

13. Go over the entire surface of the lid and knob using a range of abrasive grits. As a final surface preparation, use (0000) steel wool. Remove the lid from the glue disc by following the discussion in task 7 regarding the base and its glue disc. You need to be especially careful when you are handling the lid because of its thin, tapered edge. Then bring the tenon surface to finishing readiness by using abrasive paper and steel wool.

14. Test the lid for fit in the base and its recess. If the fit is too tight, carefully sand the tenon or lid edge using a fine-grit abrasive paper. Then refer to Chapter 43 for finishing procedures.

22
LARGE BOX WITH LID

Andy Matoesian is fond of turning this box design. It's interesting to note that this dedicated cutter resorts to the use of scrapers when turning the box base. However, he also uses a deep bowl gouge with this project.

If you compare the design in Illus. 80 with the design in Illus. 77, you can see that the overall appearance of a box can be significantly changed by altering the knob design or the thickness of the base. Andy likes to vary knob designs to make each box unique. By reducing the thickness of the base from 2″ (8/4's) (the design in Illus. 77) to 1½″ (6/4's), he's been able to create the flying-saucer type of box design in Illus. 80.

Andy generally does some paper-and-pencil design work before he begins a project, and especially when he uses thick stock of considerable diameter. Designing knobs can be great fun, particularly when you plan to spindle-turn them.

The drawing of the base, lid, and knob in

Illus.80. Matoesian boxes with lids

Illus. 81 should not only give you an idea of the overall shape of the design, but it should also serve as a guide for turning.

Although you can turn the base using either double-face tape or a glue disc, Andy prefers using a six-in-one chuck. Unless you've developed some confidence in the use of double-face tape, you should probably use the chuck or a glue disc. The base block for the box should be 2″ (8/4's) thick and from 7″ to 8″ in diameter. *With a block this size, you need to be especially concerned with the issue of safety when you turn it.* For purposes of demonstrating this project, I'll be using a glue disc. You may want to refer to the discussions in Chapter 1 on the preparation and use of glue discs for faceplate turning.

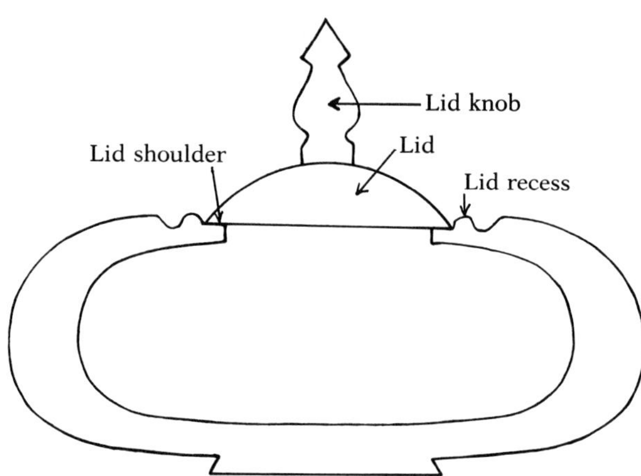

Illus.81. Base and lid

TASKS:

1. Depending on which base thickness you prefer, prepare a block from either 1½″ (6/4's)- or 2″ (8/4's)-thick stock. To help you decide on the thickness, you may want to compare the boxes in Illus. 80 and Illus. 77 once more. Now pattern and cut the base block to a diameter of at least 7″. Unless you have a commercial chuck, you should use a glue disc and a 3″ faceplate. At this point, you may want to review the discussion about glue discs in Chapter 1. Andy tends to turn this box design at 900 to 1200 rpm. Now mount the assembly on the lathe.

2. For some assistance in turning the base, refer to the specific reference guide shown in Illus. 82. As an initial task, Andy cuts the bottom tenon ¼″ long and approximately 5½″ in diameter. The tenon elevates the box and greatly enhances its overall appearance (see A 1 in Illus. 82). However, you may or may not want to include the tenon, depending on your own design preference.

Using either a bowl gouge or a 1″ roundnose scraper, round over two points (A 2 and A 3). As you true the block with the tool, also shape it at these points. To obtain a neat, clean area where the tenon joins the base (A 4), use either a diamond-point tool or a square-nose gouge.

3. Prior to turning the front surface of the base block, Andy aligns the tool rest in front of the block. He also increases the lathe speed to 2000 or 2400 rpm. To cut the bead at point A 5, Andy uses a "cheater." You may want to refer back to Chapter 3 for a description of this homemade tool and how to use it. You may also want to refer back to Illus. 24 to see how to make the cheater. Using a cheater or a tool of your own preference, make the bead at point A 5.

To make the inclined shoulder at point A 6, Andy likes to use a diamond-point tool. By making very light cuts with the diamond-point, you can make a V-shaped indentation that is approximately 3⁄16″ deep. Be sure the diamond-point is sharp so that you can get nice, clean surfaces on both sides of the V.

4. The lid-and-storage opening in the base should be anywhere from 3″ to 5″ in diameter. The size of the opening depends entirely on your own design preference. Make a cut at point A 7 with either a ½″ square-nose scraper or a parting tool This cut is for the lid shoulder and should be approximately ⅛″ wide. The shoulder should also be recessed ⅛″.

To remove wood from the storage area, Andy uses a heavy, homemade ½″ square-nose scraper (Illus. 83). You should remove a considerable amount of wood with this scraper, but leave a bottom thickness of approximately ¼″.

For removing wood from point A 8, Andy prefers to use a ¼″ roundnose scraper, which has been freshly ground with a large burr left in place. You should place the scraper at an angle inside the base so that it will follow the external contour of the base. Adjust the tool rest to minimize tool overhang as you turn the inside area.

5. After he has hollowed out the inside of the

base, Andy uses a ⅜″ roundnose scraper that is extremely sharp. The scraper is tipped with tungsten, and you can use it with either metal or wood. You should clean up and smooth the interior of the base by making very light cuts, which have a burnishing effect on the wood's surface.

6. Once you've turned the base of the box, it's time to sand its various surfaces in preparation for finishing. *At high speeds, sanding the inside surface of such a large base can be dangerous; so Andy wears a leather glove to prevent his fingers from getting cut or broken. Even if you wear a glove, Andy feels you should still approach this task with great caution.* While you can use any of the available abrasive papers, Andy likes to use a commercial sanding screen called Wetordry Fabricut. Most auto body shops use this type of abrasive, and may be willing to sell you a few pieces. Andy prefers using a 150-grit sanding screen, which is balled up in order to fit the contour of the interior of the base.

Once you've sanded all the surfaces of the base, remove the assembly from the lathe. Andy likes to cut the turned base from the glue disc with a band saw. You may want to refer to Chapter 1 for a discussion of this procedure.

7. You should pattern the lid block from stock that is at least 1″ (4/4's) thick, and you should determine the diameter of the lid block by the final diameter of the lid-recess area in the base. But pattern and cut the rough lid block at least ½″ larger than the recess area. This will give you a sufficient amount of wood for truing up the block.

8. For turning the lid, Andy uses a metal screw chuck with a screw that extends about ½″ from the chuck's front surface, but you may prefer using the wooden screw chuck described in Chapter 1. Drill an undersized hole to a depth of approximately ½″. This hole will fit tightly on the chuck screw. Andy's chuck requires a ⅜″-diameter hole; so subsequent procedural references will refer to this diameter.

Mount the block on the chuck and true it up using a roundnose scraper. The front surface of the block will be the bottom of the lid. Thus, you can turn the surface either flat or concave, depending on your own preference. You can also cut a tenon to an appropriate diameter that will fit inside the storage area of the base.

With the base surface turned, secure a ⅜″-diameter brad-point bit in a Jacob's chuck that is secured in the tailstock. Drill a hole into the block to meet the ½″ hole that was drilled earlier. You will need this ⅜″-diameter hole through the lid in order to hold the knob tenon.

Reverse the lid block on the chuck (Illus. 84). Using a roundnose scraper, round the lid at

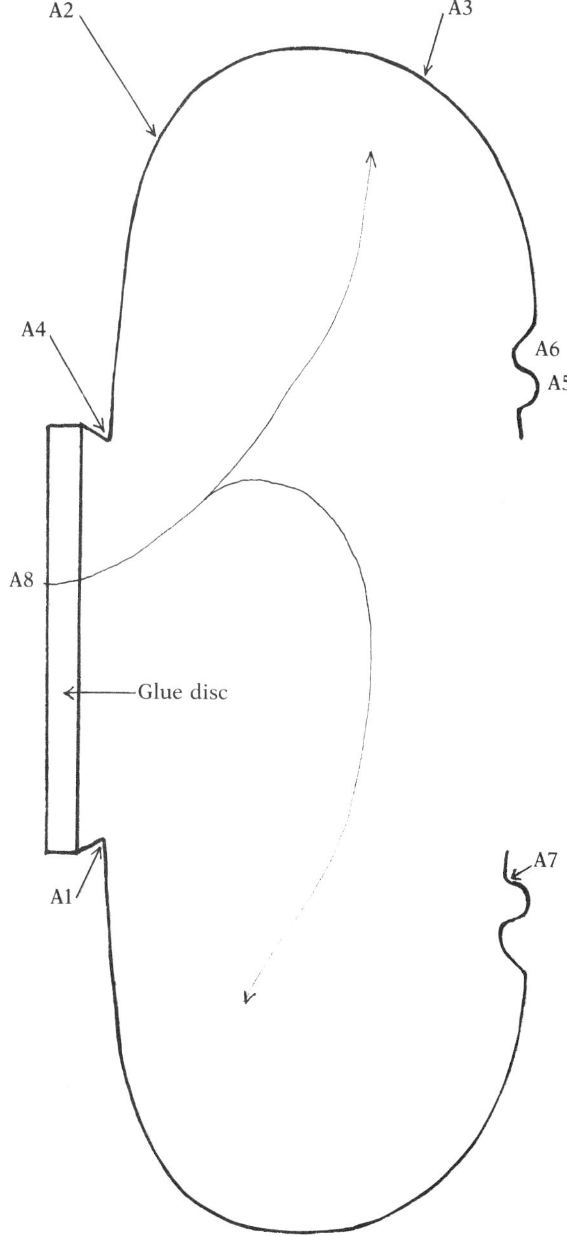

Illus.82. Base schematic

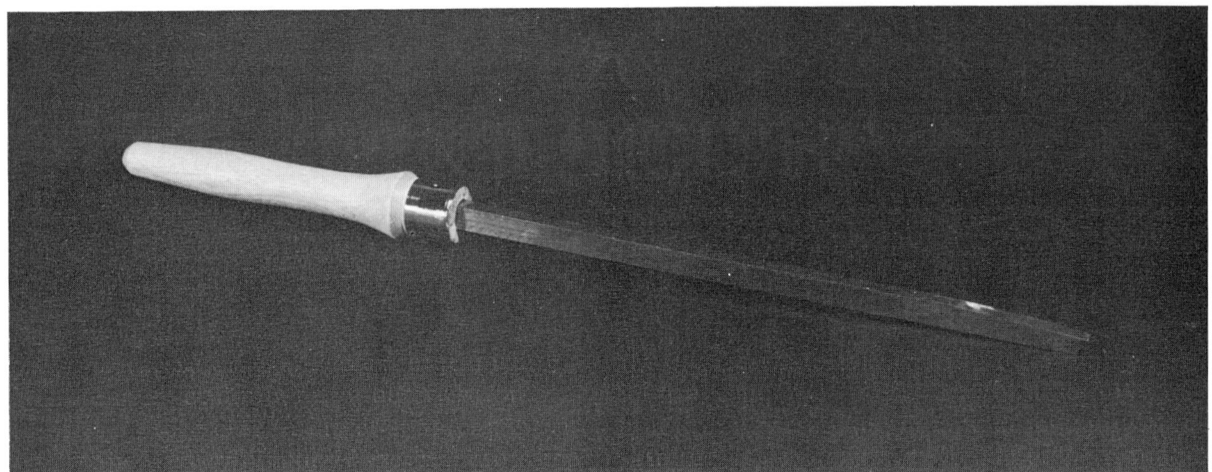

Illus.83. Homemade square-nose scraper

point B 1. The lid should be about ½″ in the middle and taper down to a very thin edge. Andy tends to roll the tapered edge so that it matches the bead, which was cut into the surface of the base with the cheater.

Rather than using a vernier caliper, Andy

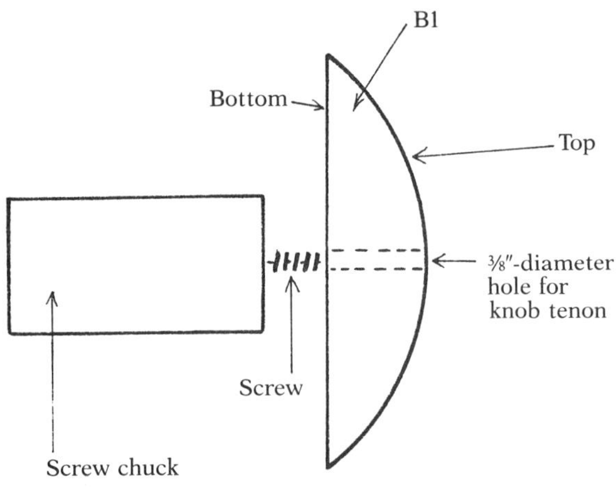

Illus.84. Lid block on chuck

prefers to turn the lid to an approximate diameter and then check it for fit on the base. Because of the deep threads in the screw chuck, you can easily remove the block and then reattach it without making it off-center. Although this is an "eyeball" procedure, Andy finds it effective for turning the lid to the precise diameter needed to fit inside the recess.

9. When it comes to turning the knob for the lid, Andy often becomes animated as he begins to discuss all the possible options. No doubt this is because the knob involves spindle turning. Also, the way the knob is turned can greatly affect the overall appearance of the box.

For assistance in turning a sample knob, refer to Illus. 85. Prepare a knob blank that is approximately 1¼″ square and 5″ long. The wood grain should be running lengthwise, or parallel to the lathe bed. Mark the ends of the blank, using diagonal lines to locate the middle. Andy uses a minidrive center on the headstock and a revolving center in the tailstock to turn the knob. Now secure the blank between centers.

10. You should round the blank using a roughing-out gouge. With a parting tool, mark the length of the knob tenon, which should be slightly longer than the thickness of the lid. With the parting tool, reduce the tenon at point C 3 (in Illus. 85) to a diameter of ⅜″. You should use a vernier caliper when you are turning the tenon to diameter.

After completing this procedure, cut a shoulder or partial cove at points C 2 and C 3. Andy really gets excited about cutting a shoulder or a cove. I don't know if the excitement is over the technique of cutting the shoulder with a ¼″ gouge instead of a parting tool or whether it's over the final design that's emerging. Andy applies the ¼″ gouge on its side and at an angle (Illus. 86). The main problem with this procedure is that the bevel of the gouge doesn't have any wood to rest on so that, in effect, it's suspended in midair before it starts to cut. However, if you do the procedure properly, you will

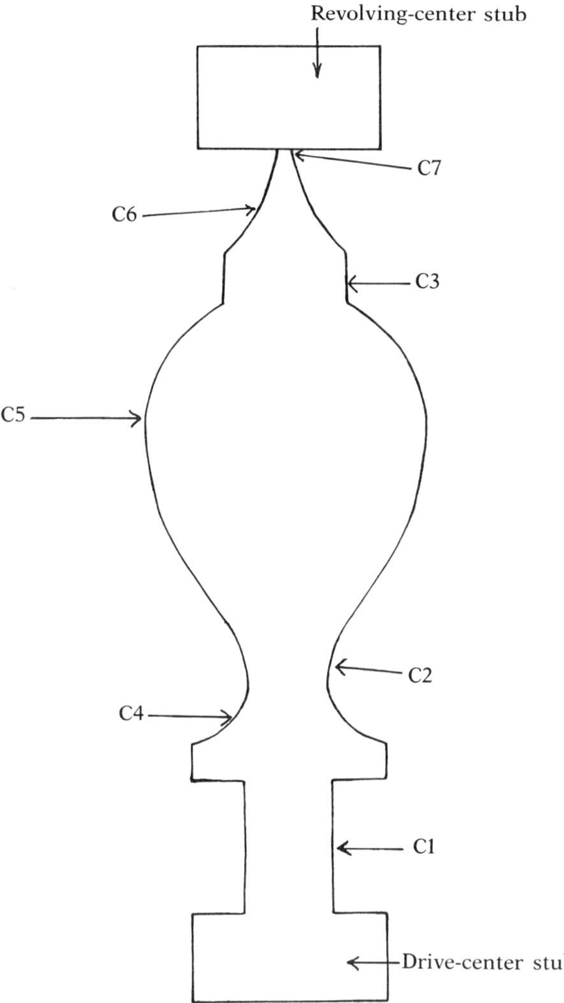

Illus.85. Lid knob schematic

get a silky, smooth, and sharp shoulder that does not require sanding.

Make your next cut at point C 4 with the ¼″ gouge. This cut leads naturally into making the bead at point C 5. With the shoulder at point C 3 cut, pare down the spindle at point C 6. Make the final cut at point C 7 with the gouge on its side. As the spindle separates from the stub, try to catch it with your left hand. Incidentally, no abrasive papers are required; the surface is ready for finishing.

11. Test the knob tenon for fit in the drilled hole in the lid. Although you can simply glue the knob tenon into the ⅜″-diameter hole in the lid, Andy tends to dress up the procedure. On occasion, he will thread the tenon and tap the lid hole in order to create a removable knob. He also will cut the knob tenon on the band saw and make a small wedge from a decorative piece of wood, which he drives into the cut; this gives the bottom surface of the lid a very interesting effect.

Regardless of how you decide to secure the knob tenon in the lid, you should sand off any excess wood from the tenon that may project beyond the lid's bottom surface. Then turn to Chapter 43 for finishing procedures.

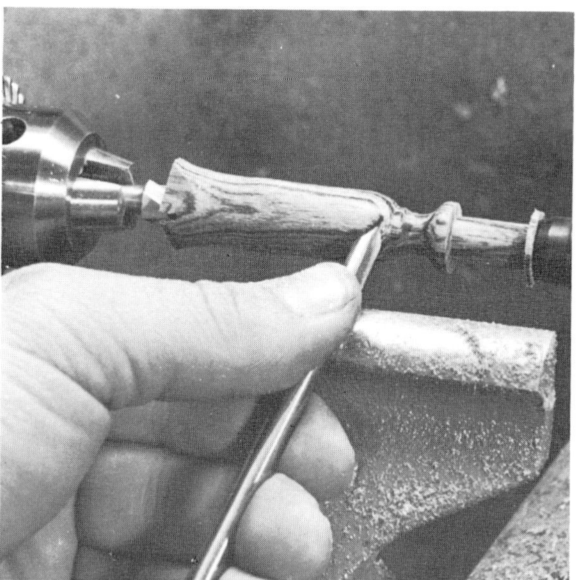

Illus.86. Lid knob being turned with ¼″ gouge

23
BOX WITH INLAID LID OF CAST & SHAPED PEWTER

With this decorative and functional box project, you get the opportunity to do some creative inlay work on the lid. In recent years, I've become fascinated with inlaying lids on turned and joined boxes. While I've done inlay work using a range of exotic woods and veneers, metal and stone are materials that currently hold my interest.

Illus.87. Box and lid with pewter inlay

As you begin exploring ways to embellish the lid or base of a box, you will be amazed at all the possible designs you can create. You will also be surprised by the variety of materials that can be used for making inlays. In addition to metals and stone, I've also been developing some designs made with various types of colored glass. Embellishing boxes can be a fascinating process and one that is limited only by your imagination and your willingness to try something different.

Inlaying a box lid not only enhances the over-

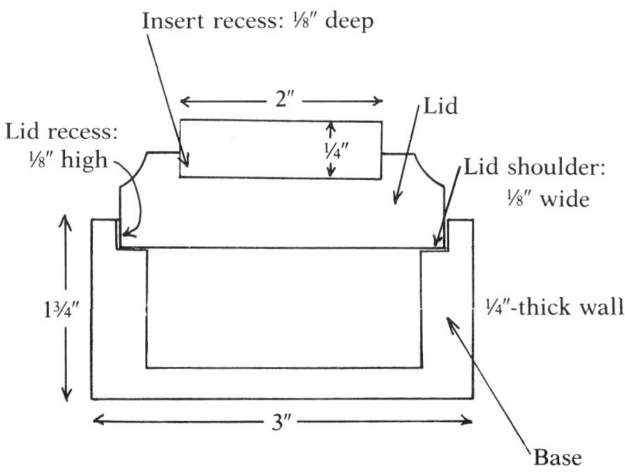

Illus.88. Base and lid

Illus.89. Moto-Tool and router attachment

all appearance of a project, but it's also an excellent way to salvage wood that otherwise might be worthless. Many of the lids that I use for small boxes have inlays placed over checks, knotholes, or rot, or some other flaw in the wood. Therefore, you may want to look through your scrap box for material for this project before cutting into good stock.

As a rule, I turn the base of the box from 1¾" (7/4's)-thick stock. The diameter of the base block is usually about 3¼". I cut the lid block from 1" (4/4's)-thick stock, rough-patterned to a 3¼" diameter. Illus. 88 shows a drawing of the box, along with some approximate dimensions. The lid has a recess cut into its top surface where an inlay insert is placed; the actual inlay is then set into the insert. This will become more apparent shortly.

Whenever you do inlay work, some special tools are required. A small Dremel Moto-Tool with a router attachment can be particularly useful (Illus. 89). Small router bits are available for these Moto-Tools, which will allow you to do some detailed routing. If you don't have a Flex-Shafts tool, you can use the Moto-Tool along with a mandrel and small sanding cylinders to shape the inlays and the lid-inlay inserts. You can also use any number of other sanding devices for shaping inlays. Additional information on tools, supplies, and procedures are included with the various tasks.

Pewter is readily available from a number of metal suppliers. I make a point of ordering a small quantity of pewter in the least expensive form. Pewter comes in sheets, discs, ingots, and cutoff pieces. You will quickly learn about prices and quantity and the best type to order from your supplier. I like to use pewter for inlays because it has a relatively low melting point. You can melt pewter in a small crucible using a propane torch or by placing the crucible in a wood stove (Illus. 90). After the pewter

Illus.90. Crucible, pewter, and tongs

has melted, pour it from the crucible into a precut wooden mould. Then shape the pewter casting using small sanding cylinders. These procedures will be discussed in detail as we proceed with the project. *Incidentally, it's important to remember that pewter contains lead; so you should wear an appropriate mask when working with it.*

TASKS:

1. Prepare a base block that is 1¾″ (7/4's) thick and has a diameter of at least 3¼″. You should also prepare a lid block from 1″ (4/4's)-thick stock with a 3¼″ diameter. Then attach a 3″-diameter faceplate to the base block and mount the assembly on the lathe.

2. True the base block with a 1″ roundnose scraper. Then, with the same tool, turn the outer wall either slightly concave or perfectly straight; the shape of the base is up to you.

3. Align the tool rest in front of the base block. Prior to removing wood from the storage area in the base, note the lid recess and shoulder in Illus. 88. The recess should be at least ⅛″ deep and the shoulder should be ⅛″ wide. I cut both of these areas using a ½″ square-nose scraper, which is usually adequate for making these cuts and also for removing the balance of the wood from the storage area. Now make the necessary lid cuts and then remove the wood from inside the base. Leave a bottom thickness of at least ⅜″, which should prevent you from hitting the penetrating faceplate screws, if you used short screws.

4. Starting with 100-grit abrasive paper, sand the inside and outside of the base. Be careful not to reduce the recess and shoulder. Follow the 100-grit paper with a series of finer grits. Then prepare the surfaces for finishing. You may want to go over all the surfaces with (0000) steel wool. Then remove the assembly from the lathe. Also remove the faceplate from the base. Now fill the faceplate-screw holes with either a wood filler or small plugs. You may want to refer to Chapter 1 for a brief discussion on some of the ways to deal with screw holes.

5. For turning the lid block, you will need to use a screw chuck. Illus. 6 shows two homemade chucks that you can use for turning box lids. I prefer using the shorter chuck for this procedure. If you don't have a screw chuck, you may want to review Chapter 1 for specifics on how to make one. Of course, you can always purchase a commercial chuck or secure the lid block to a pine disc using double-face tape or white glue.

If you are using a screw chuck, drill an undersized hole through the middle of the lid block. You want the lag screw to thread tightly through the drilled hole. Be sure the drilled hole is in the middle of the lid, or you will be turning off-center and removing too much wood.

6. With the screw chuck on the headstock, thread the lid block onto the lag screw. Before you begin truing up the block, determine the diameter required for the lid using a vernier caliper. Set the short prongs of the caliper by placing them against the wall of the lid recess. This automatically sets the longer prongs to the required lid diameter.

Using a ½″ roundnose scraper, true up the lid block and turn it to the required diameter. During these procedures, keep the wall of the block square. Also, be certain to monitor the diameter of the block with the caliper as you turn. The final lid diameter should be a fraction smaller than needed; this way, you will get a comfortable-fitting lid rather than one that's too tight.

7. In Illus. 88, you can see how the top portion of the lid block is turned inward and slightly concave. The concave taper directs the eye towards the inlay-recess insert and the inlay proper. Since this cut is made strictly for decorative purposes, you may prefer decorating the side of the lid another way. I make the concave cut using a ½″ roundnose scraper, but you can also use a ¼″ fingernail gouge.

8. Align the tool rest in front of the lid block. It's necessary to cut the inlay recess into the top surface of the lid. As a rule, I make the inlay recess about ⅛″ deep with a 2″ diameter. You will find it necessary to turn along the side of the lag screw. The small tower that develops around the lag screw can be removed later with a wood chisel. A wall that is about ¼″ wide should surround the recessed area. With a light touch of the scraper, clean up the surface of the wall. Then sand the surfaces of the lid using a range of abrasive grits, but be sure to leave the top edge of the recess sharp.

9. Remove the turned lid from the screw chuck. With a wood chisel, cut off the small tower that developed around the lag screw. Fill the lag-screw hole at the bottom surface of the lid with wood filler. Then test the lid for fit on the base. At some point, remember to sand the bottom

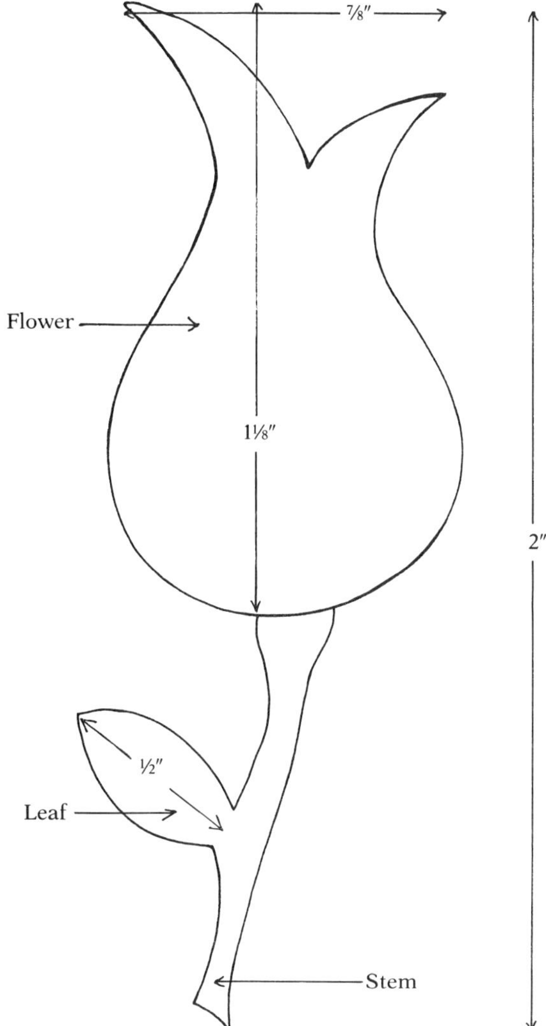

Illus.91. Flower, stem, and leaf inlay

surfaces of both the lid and the base to finishing readiness.

10. Prepare the inlay-recess from a wood that contrasts with the wood you used for the lid. For example, in Illus. 87, the base and lid of the box are made from black walnut and the insert is made from Indian padouk. When you are selecting wood for the insert, keep in mind that its color should also set off the pewter inlay.

With a ⅛″-deep insert recess in the lid, you should cut the inlay-recess insert to a width of at least ¼″. You want some of the insert to extend over the top surface of the lid. This thickness also tends to better showcase the actual inlay when it's in place. Using a plastic template or school compass, pattern a 2″-diameter insert and cut. Then place the insert into the recess to check for fit. It's best if you have a tight fit because the insert's diameter will get somewhat reduced from later procedures.

11. This task involves designing and preparing the pewter inlay. I'm especially fond of flower inlays, and they're both easy and fun to make. You may want to refer to Illus. 91 to note an inlay pattern of a flower, stem, and leaf. The pattern includes some approximate dimensions to assist you in your planning activities. If you prefer creating your own design for an inlay, do some pencil-and-paper design work. Remember that the insert diameter is only 2″. For a different effect, you can extend a portion of the inlay onto the surface of the lid. Part of the lid surface would then need to be cut out to accommodate the portion of the inlay.

For casting the pewter into the inlay design, I use a piece of hard maple. Trace or draw freehand the inlay design on the surface of the hard maple board. Using a Dremel Moto-Tool with a router attachment and a small router bit, rout out the design in the board. With a ¼″-thick inlay insert, you will want the pewter inlay to be at least 5/16″ thick. By making several passes at different bit settings, you can rout the inlay design into the board.

After you have routed the inlay design into the hard maple board, it will be time to melt some pewter. Place a sufficient amount of pewter in a small crucible. Have a pair of tongs ready for holding the crucible during the melting process and also for pouring the pewter into the routed-inlay design (Illus. 90). As indicated earlier, you can melt the pewter with a propane torch or in a bed of coals in a wood-burning stove. *Remember, you're handling extremely hot metal; so proceed with caution. I always wear a pair of welder's gloves when I'm heating and pouring pewter.*

Heat the base of the crucible until the pewter melts. With the tongs appropriately placed around the crucible, pour the melted pewter into the routed inlay in the hard maple board. Try to control the amount of pewter you pour so that you don't overfill the routed area. Using a small sliver of wood, push the pewter around, forcing it into all the routed areas of the design.

The hot pewter will char the surface and edge around the routed-inlay area, but this is not a problem. I've used the same routed-inlay design at least five times before the charring destroyed the recess in the wood. Allow the pewter to cool for at least 30 minutes. *After that time, the pewter may still be hot; so be careful when you touch it.* To remove the inlay from the board, turn the board upside down and tap it on a concrete floor; then the inlay should drop out of the routed area.

12. When the inlay is cool enough to handle, you need to shape it using small sanding cylinders on a Dremel Moto-Tool or a Flex-Shafts tool. You may have some other type of sanding cylinder that will work just as well. *To avoid breathing in the lead in the pewter, you would be wise to wear some type of protective mask.*

Shape the flower, leaf, and stem so that they resemble real ones. For example, you can round the bottom part of the flower to make the flower appear to be lying on some surface. You can make the top surface of the leaf slightly concave. And you can make the stem appear rough and varying in thickness. Use your imagination in this enjoyable process. In addition to using sanding cylinders, I also use small carbide or high-speed steel burrs to make the details on the inlays.

After you have shaped and detailed the inlay to your satisfaction, you may want to polish it. Almost any polishing compound and a cloth buffing wheel will make the pewter shine. However, I prefer the dull appearance of the pewter; so I generally don't polish pewter inlays.

13. Place the inlay on the top surface of the inlay insert. Leave some wood between the top edge of the flower and the side of the insert. If this results in the stem extending beyond the edge of the insert, you can cut an area in the lid's surface to accommodate the extending stem. Using a sharp pencil, trace an outline of the inlay onto the surface of the insert.

With either a scroll saw or a band saw with a ⅛"-wide blade, cut out the traced inlay outline from the insert. It's best to cut on the inside of the tracing lines. While you're cutting along the traced lines, you may prefer to cut the insert into several pieces (Illus. 92). This procedure makes shaping the inlay insert considerably easier. However, either procedure is acceptable for preparing the inlay insert to hold the inlay in place.

Using a Moto-Tool or Flex-Shafts tool and sanding cylinders, shape the insert parts. Round the insert edges to create an interesting effect. Then reduce the edges that are next to the inlay. This will set off the inlay better when it's in place.

14. Place the insert parts and the inlay inside the lid recess. If the fit is too tight, reduce the outer edge of the insert using a Moto-Tool and a sanding cylinder and then reassemble the insert parts in the recess. If the stem extends over the surface of the lid, trace its outline onto the lid surface with a sharp pencil. Rout out the area using a Dremel Moto-Tool with a router attachment and small bit. The depth of this area should be the same as the lid recess—⅛". Replace the insert parts and the inlay to check for fit.

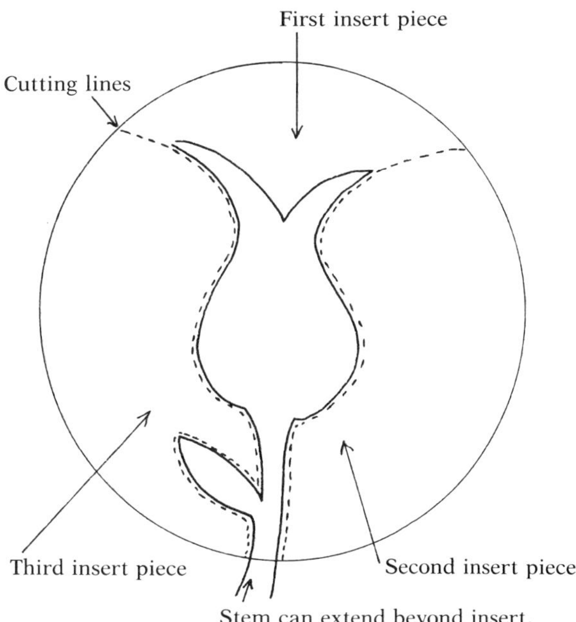

Illus.92. Insert cutting lines for inlay

After removing all the parts from the lid recess, spread wood glue over the bottom surface of the recess. If you've routed a stem area into the lid's surface, spread some glue in this area as well. Place the insert parts and the inlay into the recess and then press them into the glue. Allow the glue to dry. Then refer to Chapter 43 for finishing procedures.

24
BOX WITH INLAID LID OF SHAPED ALABASTER

This box project is basically the same as the one presented in the previous chapter. The only difference is the type of material used for the inlay in the lid. Whereas the previous project employs pewter for the lid inlay, this project uses stone.

In my search to find materials to use as inlays, I've discovered alabaster and soapstone (Illus. 94). For this box project, I've used alabaster, but you can basically use the same procedures with soapstone.

Alabaster is a translucent, crystallike material that easily lends itself to inlay work. The stone seems to have few imperfections in it, and can be rubbed to a high gloss. Best of all, I've found that alabaster can be cut with both a band saw and a scroll saw. With these standard woodworking tools, I've been able to prepare alabaster inlays in a variety of intricate designs.

Soapstone is a rather soft stone that is a beautiful sea-green color. Although you can't cut it very easily with a standard band-saw blade, you can certainly shape it with burrs and sanding cylinders. To obtain pieces of soapstone for inlays, I generally break off what I need from a large piece with a hammer, but you can also cut soapstone with metal-cutting equipment.

All the procedures for making the base and lid for this project are the same as those presented in Chapter 23. The inlay insert for the lid should also be prepared according to the procedures and dimensions suggested in Chapter 23. You will trace the inlay onto the surface of the insert, which you will then cut into pieces to hold and showcase the inlay.

Illus.93. Box and lid with stone inlay

Illus.94. (Left) soapstone and (right) alabaster

TASKS:

1. It is assumed that you have turned the box and prepared the inlay insert for the lid according to the directions presented in Chapter 23. Remember to use a type of wood for the inlay insert that will contrast well with the wood used for the lid and set off the actual inlay. Indian padouk, for example, is an excellent choice of wood to use for the insert when you're using an alabaster inlay. To minimize wasting inlay material, the inlay insert should be no thicker than ¼″. It's assumed that you have made the lid recess approximately ⅛″ deep.

2. Using a band saw and a ½″-wide blade, cut a slice of alabaster that is at least ½″ thick, which is more than adequate for the inlay when the insert is ¼″ thick. You may want to refer to Illus. 95 to see a drawing of the inlay design for this project.

Although you can make a pattern of the flower, you may prefer drawing the design directly on the stone. The flower should be at least 1″ long and approximately ⅝″ wide. If you have a scroll saw, cut the patterned flower from the ½″-thick stone. *You should wear a face mask or some other type of eye protection whenever you're working with unfamiliar hard materials.*

3. As Illus. 95 indicates, in addition to the alabaster flower, you need to prepare a stem and leaf for the inlay design. While a variety of woods can be used for the stem and leaf, I generally use poplar and walnut for this design.

Using scrap material, prepare a piece of walnut for the stem that is about ½″ thick. Then cut a piece of poplar (or maple, if poplar isn't available) to the required thickness for the leaf. On the piece of walnut, draw a stem, freehand, that is approximately ½″ long and ⅛″ wide; on the piece of poplar, draw a ⅜″-long leaf. Now cut out the stem and leaf using a scroll saw or another saw that's similar.

4. Place the rough stone flower on the inlay insert and trace its outline using a sharp pencil. Place the walnut stem at the bottom edge of the flower's outline and trace it. (You may want to refer to Illus. 95 again to note how the various parts are placed on the insert.) Then set the leaf next to the stem and trace its outline onto the surface of the insert.

Cut out the traced areas on the insert using a scroll saw. As Illus. 95 indicates, you can cut the insert into three pieces, which makes it more interesting in appearance and also easier to shape. Cutting the insert into three pieces makes assembling the inlay and insert back into the lid recess much easier as well.

5. Using either a small Moto-Tool or Flex-Shafts tool and sanding cylinders, shape the various inlay and insert pieces as desired. If you have small carving burrs, you may want to use them for shaping the stone flower. Make the flower, stem, and leaf as realistic-looking as possible. You will need to reduce the thickness of the stem and leaf so that they look proportional to the flower. Picture the way a stem joins the bottom of a flower and then shape the parts accordingly. Do not reduce the width of any of the parts or you will have gaps in the inlay.

The various edges of the insert pieces should be slightly rounded, especially those edges that are next to the inlay parts. The purpose of the insert is to show off the inlay; so shape the insert pieces accordingly.

6. When you have shaped all the pieces, assemble them in the lid recess to test for fit. If the fit is acceptable, remove the pieces and spread wood glue on the bottom surface of the recess. Assemble the inlay and the insert pieces in the recess and then press them into the glue. Allow the lid to dry. Then turn to Chapter 43 for finishing procedures.

By the way, if you are unable to obtain alabaster and you want something similar for white inlays, you may want to try Tagua palm nuts. These nuts are a vegetable ivory and they're an excellent material to use for inlay work.

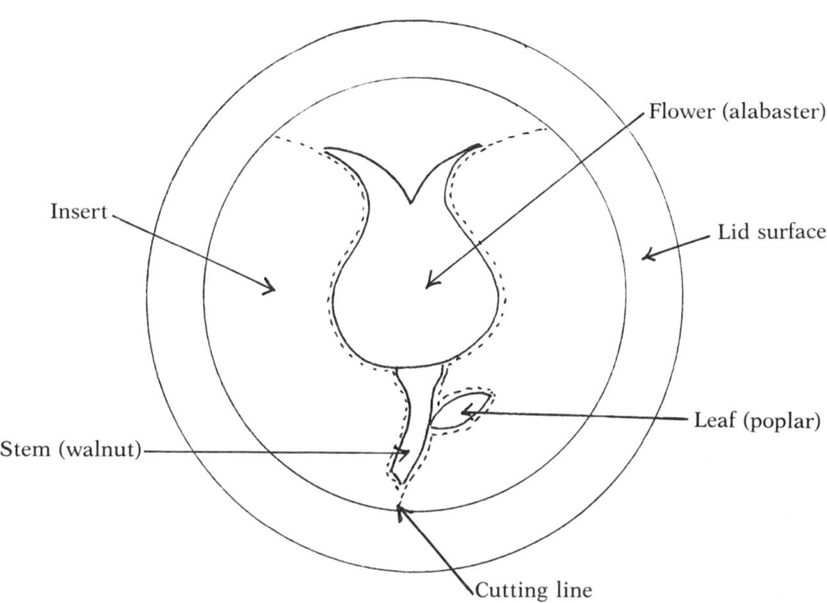

Illus.95. Lid inlay

25
TOOTHPICK HOLDERS

This project is simply a small turned piece, which you can use for holding toothpicks, wooden matches, or other small items. You can make these small containers from your thicker scraps of wood. While the far left and the far right containers in Illus. 96 have been turned from bocote from the cordia tree, the others have been turned from scrap material.

Although size is not a major consideration with this type of project and is primarily dictated by available scraps, I try to make the containers from 1½″ (6/4's) (or thicker) stock. Toothpicks are generally about 2⅝″ long and wooden matches are usually about 2¼″ long; so wood that is at least 1½″ thick is ideal for these containers. The finished diameters of the containers range from 1½″ to 3″, depending on the availability of scraps. If you want to use this project for decorative purposes only, these dimensions will be somewhat irrelevant.

It's best to turn this project using either double-face tape or white glue and a pine disc.

Illus.96. Toothpick holders

Then you mount the assembly on a 3″ faceplate and turn using standard-size scrapers. You may want to refer to the discussion on using tape and white glue in Chapter 1. Given the size of the containers, I generally use double-face tape because it is simpler, less of a mess, and more efficient.

If you prefer spindle turning, you can prepare blanks, with the grain, to mount between centers. Given the diameter of the blanks, it's best to use a standard four-pronged drive center. Although you can use a dead center in the tailstock, I prefer to use a revolving center. You can drill the storage area later on a drill press.

I often use a 1″ (or larger) multispur bit to drill the storage area. While you can use the tailstock to drill blocks secured to a faceplate, I usually drill the holes using a drill press. On some of the more decorative designs, I turn out the storage area, often rounding the bottom surface. For example, the two bocote containers in Illus. 96 both have rounded bottoms. Because of the depth of the storage area, the round bottom won't prevent you from using these containers for holding toothpicks or matches.

With these small containers it's a good idea to do some pencil-and-paper design work. Whether you plan on using the faceplate or spindle-turning method, try working out a design before you begin turning. This procedure not only results in more interesting containers, but also helps you to further develop your design and turning skills.

When you have completed the project, turn to Chapter 43 for finishing procedures.

26
PENCIL HOLDERS

Desk-top accessories are excellent turning projects. Although a number of other desk-top accessories are presented in later chapters, these pencil holders are included here because they are an extension of the toothpick-holder design presented in the previous chapter. They are simply turned from more wood in order to accommodate larger items in their storage areas.

In Illus. 97, you can see two basic designs and types of wood that can be used for this project. The pencil holder on the right was turned, using a faceplate, with the grain running parallel to the lathe bed. This type of design is suitable for most contemporary desks and office settings. The other pencil holder was made from a scrap piece of Osage orange (hedge), which had been cut from an old fence post. It, too, was faceplate-turned, but still retains the natural character of the wood. No doubt this pencil holder would be totally out of place in a streamlined-office setting. Desk-top accessories often reflect the style or personality of the indi-

Illus.97. Pencil holders

vidual using the desk. As you plan this project and select your wood, keep in mind the user and the potential setting.

Since standard pencils are approximately 7½" long, the pencil holders should be at least 3½" high. Although this isn't very high, the pencils will fall against the inside wall but will not tumble out. Unless you have very thick stock, you may want to consider turning the pencil holders with the grain running parallel to the lathe. As a rule, the pencil holders should be at least 2½" in diameter. If the diameter is smaller on a container this high, the container will tend to tip over when items are placed in it. You should do some design work with this project and take into account both the height and diameter of the pencil holder.

You can spindle-turn these containers and drill their storage areas using a large multispur bit. If you cut a tenon into the bottom edge of the container, you can also mount it into a three-jaw chuck for turning the inside storage area. My own preference is to turn the pencil holders using a faceplate and a glue disc. Given the length of the block, I prefer using glue rather than double-face tape.

For some assistance in mounting blocks when using a faceplate, refer back to Chapter 1. When you have completed the project, refer to Chapter 43 for finishing procedures.

27
SMALL BOWLS

Bowls of any size are great fun to turn on the lathe. While turning large bowls can be somewhat more challenging, I find turning small ones to be more enjoyable. By small, I mean bowls that have a diameter of 3″ to 4″ at most. They can range in height from 2″ to 4″. Best of all, you can usually turn small bowls from scrap material.

The bowls in Illus. 98 were turned from scraps, some of which came from my backyard woodpile and had been drying for a number of years. Except for a small piece of hedge fence post, I cut the turning blocks from pieces of wild cherry, black locust, and white ash. Most of the logs had been split; so it was relatively easy to select pieces that had some character.

While you may not be so fortunate as to have such a variety of hardwoods in your woodpile, I'm confident you have indigenous woods in your area that are ideal for turning small

Illus.98. Small bowls

bowls. Also, you probably have chunks of wood around your shop that can be used for small bowls. How about those pieces that you thought were too flawed to use and so set aside someplace in your shop? Some years back I turned some beautiful small bowls from a pallet, which had been used for packing a large tool. The tool was manufactured in Brazil, and the pallet that supported it during shipping was some type of local wood. I considered it an exciting find and immediately turned the unfamiliar wood into small bowls. A wood-crazed turner can find wood for projects in all kinds of amazing places.

It's best to turn the small bowls in this project using a 3″ faceplate and a 3½″ glue disc. Although you can use double-face tape with blocks this size, I'm inclined to use a pine disc and white glue. You may want to review the discussion in Chapter 1 on securing blocks for faceplate turning.

I prefer turning the small bowls using either a 1″ or ½″ roundnose scraper. When ground properly, the scrapers leave a relatively good surface that requires a minimum of sanding. They are especially useful in turning the inside area of the bowl, and permit you to turn a clean, round bottom surface.

When you have completed the project, you may want to refer to Chapter 43 for finishing procedures.

28
MEDIUM-SIZED BOWLS

While the word "medium" is somewhat vague, in woodworkers' parlance it generally has more to do with the amount of wood used than with any specific dimensions. When I make medium-sized bowls, I usually turn them from spalted woods. When I work with the spalts, I don't want to waste the wood or use too much of it on a given project.

Without a doubt, the spalts are my favorite type of wood to find and turn. Fortunately, I live in an area where there are excellent hardwood forests and there is substantial moisture to nourish them. It is the moisture and those marvellous growths called fungi that make spalted wood possible. Fungi need lots of moisture to begin the decaying process in wood that results in streaks of black zone lines. These lines are what makes spalted wood so attractive and desirable to the woodworker.

The origin of the word "spalt" is not clear,

Illus.99. Medium-sized spalted bowls

especially in reference to decaying wood. It seems to be one of those words that has come into common usage in our language, but is not formally recognized in dictionaries. The word seems to be taken from the German language, where it loosely means to split, divide, or break down. Historically, the word spalt was commonly used in the wood-shingle industry in the United States. A spalt, apparently, was a slice of wood that had to be removed from a shingle. No doubt this piece of wood was often rotten or decaying.

Whatever the origin of the word may be, spalted wood is clearly the result of fungal infestation. Reducing the moisture content or drying the wood will stop the decaying process because it kills the fungi. When you find your own spalted wood, the first thing you must do is bring it inside and let it dry out.

The bowls in Illus. 99 are made from spalts, which I found in the woods and in my backyard woodpile. In recent years, hardwood dealers have begun to stock a range of spalted woods because of the growing interest in them. If you are unable to find your own spalts to dry and turn, your local dealer may be a good alternative.

As a rule, I use 3½" pine glue discs mounted on a 3" faceplate when turning spalted bowls. Using white glue, I secure the glue disc directly to the bottom of the turning block. Because spalted wood is decayed wood, it is frequently too soft for holding faceplate screws; but on occasion, I have used very long screws with a faceplate and have had success in turning. The way you secure the block for turning primarily depends on the condition of the wood.

The major problem in turning spalted wood is the cross grain or mouse that tends to develop. The more decayed the wood, the more problems you will have with the cross grain. Another problem with the spalts is that you tend to pull out pieces of decayed wood with your tools. This chipping seems to occur even when you use sharp tools.

I've found that one of the best ways of dealing with both the mouse and the chipping is to use abrasive paper. After I've turned the bowl, I will go over the surfaces with 80-grit abrasive paper. If the mouse is really bad, I'll stop the lathe and go over it with the abrasive. Another effective procedure is to reverse the rotation of the block. You will, of course, need a reversing drum switch on your lathe for this procedure. It also helps to periodically move the abrasive paper across the surface at an angle. This tends to cut some of the wood fibres off, rather than to force them back into the surface.

After you've gone over the surfaces with the 80-grit abrasive paper, you should proceed with 100- and 150-grit paper. Depending on the quality of the surface that's emerging, you may also want to use a 220- or 240-grit for the final application. With some of the spalted woods, you may find that there comes a point when it's simply not worth spending any more time and energy with the sanding. Because of the fungi, some spalted woods will always remain somewhat coarse. It's best to make peace with this reality.

Although Chapter 43 presents various finishing products and procedures, it's worth noting here that spalted woods consume great quantities of finishing products. I usually finish spalted bowls with Deft, a high-grade lacquer. It's not unusual to put eight coats of lacquer on a spalted bowl. The wood is so porous that it simply soaks up the lacquer. Frequently, finishing the spalts is similar to sanding them: There comes a point when you stop applying any more products and simply enjoy the bowls as they are.

29
LARGE BOWLS

The two large bowls in Illus. 100 were turned by my son, Peter Jacobson. Peter especially likes turning bowls of this size and design. As a hobbyist woodturner, he only has a limited amount of time for turning; so he uses scrapers rather than gouges because he feels they're faster and more efficient.

Osage orange (or hedge) is his wood of choice for making large bowls. Peter used hedge for the bowl on the left in Illus. 100. This marvellous yellow wood, which some people also call bodark, is extremely hard. Its name can be traced to the Osage Indian tribe, who used the wood for making bows and other wooden items. The word "bodark" is a bastardization of the French *bois d'arc*, roughly meaning bowed wood.

Peter likes to use old hedge fence posts that have been in the ground for many years. He usually scrounges them from local farmers or nearby peckerwood sawmills. While hedge generally turns a brownish color over time, the

Illus.100. Large bowls

Illus.101. Hedge turning blocks

fence posts tend to be and remain a glorious golden color. Being in the ground only seems to enhance the color and character of the wood.

There is one problem with the fence posts: You need to watch for nails or other metal devices that might have been driven into them to secure barbed wire. You can encounter them while band-sawing a block or even, on occasion, when turning. Peter has also destroyed a number of chains while cutting through the posts with a chain saw. If you find some hedge posts to use for woodturning, be wary of the metal that may be in them (Illus. 101).

In addition to using Osage orange for bowls, Peter also likes to scavenge for bowl wood in neighborhood woodpiles. He concentrates on looking for pieces of crotch that have a lot of character. For example, he turned the bowl on the right in Illus. 100 from red-oak crotch that had begun to spalt. He was able to prepare and turn the bowl so that a portion of the bark was left on the top edge.

Peter likes to use bowl blocks that are at least 5″ (20/4's) thick and from 6″ to 7″ in diameter. While it's easy to find fence posts that meet these size requirements, finding material this size in woodpiles tends to be more difficult. There are many other kinds of wood that meet these dimensional requirements, but Peter prefers to turn hedge fence posts and woodpile stock. If you are looking for inexpensive wood for turning large bowls, you may want to consider Peter's sources of material.

TASKS:

1. Using a band saw with ¼″-wide blade, prepare a block to the previously mentioned dimensions. If your stock is large enough, you may want to make the bowl even thicker and with a greater diameter.

2. Sometimes it's necessary to sand the bottom surface of the block flat in order to attach a faceplate. This is often the case when you use wood from a fence post. You often have two rounded ends on the block. Squaring an end on the band saw can be quite risky with blocks this size. Also, hedge is a tough wood to cut. So Peter prefers squaring the bottom surface with an 80-grit belt on a 6″ × 48″ sander. Then he uses the worst surface of the block for the bottom.

Hedge, especially fence posts, can be so hard that it's almost impossible to thread a faceplate screw into the wood. Therefore, it's best to place the faceplate, centered, on the bottom surface of the block and drill small pilot holes for the screws. Peter uses a ¹⁄₁₆″-diameter bit secured in a small pneumatic drill to make the pilot holes.

You should secure the faceplate to the block using an electric drill and power bit. Peter generally uses square-recess screws that are 5⁄8″ long to secure the faceplate to the block. A screw this long is adequate for holding the block on the faceplate. Now mount the assembly on the lathe.

3. Because a hedge block is never perfectly round, Peter brings the tailstock with a revolving center forward for safety reasons. Then he secures it tightly against the block for added support. Hedge is a very heavy wood; so when a hedge block is off-center, it can move your lathe around unless it's bolted to the floor or weighed down with bags of sand. Depending on the size of the block and the extent that it can be mounted off-center, Peter sets the lathe speed at either 990 or 1475 rpm.

You should true the block and do the initial shaping with a 1″ roundnose scraper. Illus. 102 shows an outline of the design Peter wants to achieve. Notice how the bottom edge is rolled inward towards the faceplate. Now shape the bowl to its final form with a 1″ roundnose scraper.

4. Align the tool rest to the front of the block. As a rule, Peter likes to make the top hole about 2¼″ in diameter. Not only does this diameter look good, but it also permits easy access with your tool when you're removing wood. Either remove the tailstock support or use a small tool rest.

Peter makes the initial cuts in the inside area with a 1″ roundnose scraper. After removing some wood, he then finishes turning the inside using a ½″ roundnose scraper. The condition of the block determines the amount of wood you should remove from inside the bowl. Hedge fence posts often have some checks through them that can create a hazard at this phase of the turning process. If you remove too much wood, the bowl could easily break apart. *As a safety precaution, Peter always wears a full-face protector when turning these bowls.*

Peter views his bowls as decorative rather than functional; so he sees no reason to remove a lot of wood from inside. He enjoys the weight of the wood and, when he uses hedge and the treasures he finds in woodpiles, considers turning thin walls a waste of beautiful wood.

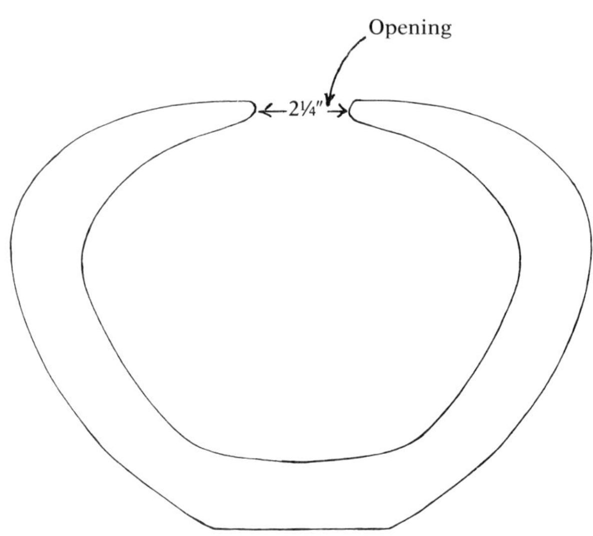

Illus.102. Outline of large bowl

5. When he has completed turning the bowl, Peter sands the outer surface, starting with 150-grit abrasive paper. Hedge is a relatively easy wood to prepare for finishing. But you should wear a dust mask when you are sanding hedge because the dust that gets generated is very heavy and can cause some mild respiratory problems. When I'm around hedge dust, my nose tends to plug up rather quickly.

After using the 150-grit abrasive paper, Peter moves up to 220-grit. Then he follows these abrasives with (0000) steel wool. A thorough rubdown with steel wool causes the golden-colored hedge to shine. Sometimes Peter doesn't sand the inside of the bowl, depending on the condition of the bowl's top hole. Often there will be a check and a sharp edge at the entry hole. When there's a sharp edge, it's too dangerous to sand the inside of the bowl; so Peter will usually do some sanding on the inside after he removes the bowl from the lathe.

6. With the faceplate removed from the bowl, Peter prepares some 3⁄8″-diameter Indian padouk plugs. He drills out the faceplate-screw holes with a 3⁄8″-diameter brad-point drill. Then he places a touch of wood glue in the holes and taps the plugs in place. When the glue is dry, he sands the plugs flush with the bottom surface and prepares this area for finishing.

When your bowl is ready for finishing, you may want to refer to Chapter 43 for a discussion of various finishing products and procedures.

30
LARGE BOWL

This project demonstrates how you can use a log that has been infested by bugs to turn a very attractive bowl. Understandably, most woodturners avoid logs that show clear evidence of insect activity. They seem concerned about the stability of the log and apprehensive about possibly bringing a swarm of insects into their shops. However, I made the large bowl in Illus. 103 from a persimmon log that had been infested with insects. When I brought the log home from the woods, I thoroughly saturated it with a general insect spray. Then I covered the exposed ends of the log with a commercial sealant, called Seal-Lite 60, and set the log in the corner of my shop. After the log had been lying in my shop for well over a year, I cut off a 4½"-long chunk and turned the bowl you see in Illus. 103.

Persimmon is a very hard and beautiful wood and one that is also very easy to turn. When you

Illus.103. Worm-infested bowl

apply a gouge or chisel to this kind of wood, you can generate large, long shavings. This makes you feel as if you must be doing something right with your tools. I'm not especially fond of the fruit produced by the persimmon tree, but, when a tree is down, it's worth going after. Because persimmon is frequently streaked with black, it's often referred to as the poor man's ebony. Whether it's in prime condition or laced with insects, it's a great wood to use for turning.

Since the insect-infested log I found was only 6″ in diameter, I decided to turn the bowl with the grain running lengthwise or parallel with the lathe bed. This is an effective way to make the most of a chunk of a log that has a rather small diameter. To keep the faceplate screws from pulling out during turning, I use screws that are at least 1″ long. It's also a good idea to place a revolving center in the tailstock and bring it forward into the log's surface for support. This type of turning can be hazardous because the faceplate screws are running parallel with the wood grain, but you also have pockets of insect activity throughout the log.

I turn this type of block using a 1″ roundnose scraper with the lathe running at 990 rpm. Inevitably your block will be somewhat off-center because of the shape of the log. Thus, you should be turning at a relatively slow speed. After partially shaping the block, I make finishing cuts to clean up the surface. I try to leave the block in its natural shape, rather than removing a lot of wood to true it up.

Align the tool rest in front of the block, while removing the tailstock-support assembly. On this type of block, I only remove a minimum of wood from the inside using a 1″ roundnose scraper. You may want to refer back to Illus. 103 to see how shallow I've turned the bowl.

Usually bowls turned from this type of wood require a minimum of sanding. Also, the small wood projections, caused by the bug infestation, can make the surfaces quite difficult to sand. Don't use steel wool on this type of bowl because the wires will be pulled out and held on the surface by the many small projections.

After you have removed the faceplate from the finished bowl, you can patch the screw holes with a natural-colored wood filler. You will find that this color matches the sawdust packings of the insects and simply makes the holes appear as more insect activity.

You may want to refer to Chapter 43 for a discussion of finishing products and procedures.

31
GAVEL & BASE

Since gavels and bases are tools of the judiciary, it's only natural that this project would be one of Judge Matoesian's favorites. Both the handle and the head of the gavel offer you an opportunity to use your spindle-turning skills and tools. These components also lend themselves to the use of an array of beads and coves. Incidentally, to add to the charm of the gavel, Andy has threaded the handle and head.

Although Andy usually rough-cuts several handle and head blanks on the band saw at one time, the tasks that follow are geared towards preparing and turning a single gavel base. It should be noted that Andy prefers to turn the gavels and bases from black walnut.

TASKS:

1. Prepare a blank for the handle that is 1½" (6/4's) square and 13" long. Even though you can turn the handle from 1¼" (5/4's)-thick stock, Andy prefers the thicker material because it offers some leeway for possible errors and any design modifications dictated by the nature of the wood.

To mark the midpoint on the ends of the

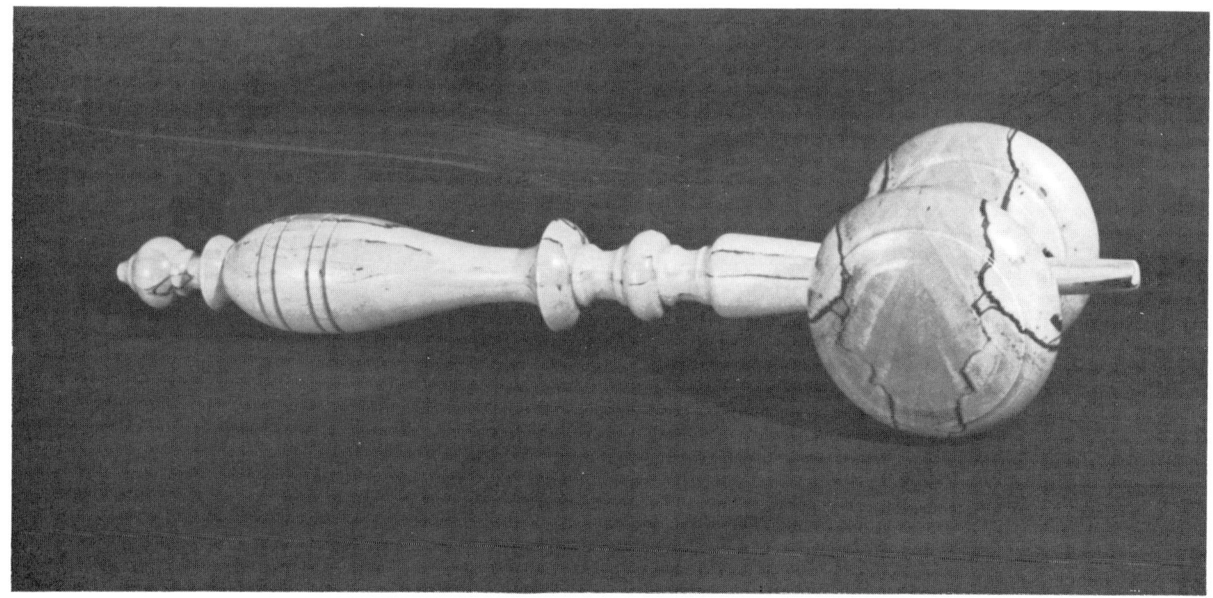

Illus.104. Gavel

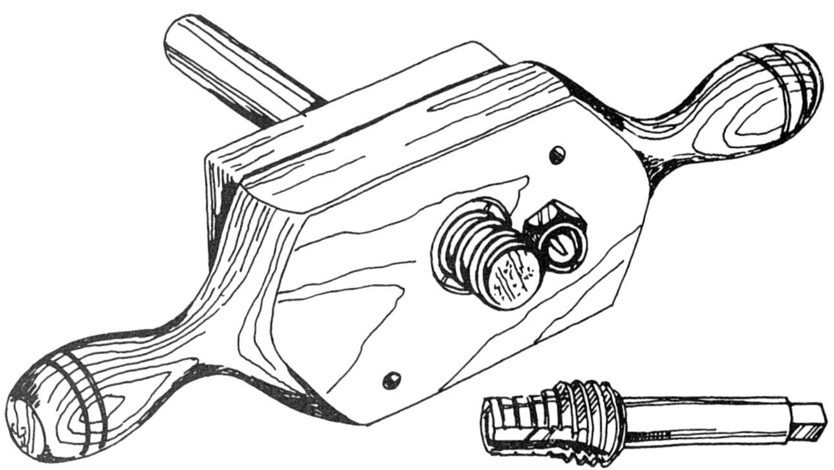

Illus.105. Threader and tap (drawing courtesy of Conover Woodcraft Specialties, Inc.)

blank, Andy draws diagonal lines from each corner. Then he drives an extra four-pronged drive center into one end. The drive center is driven in fairly deep so that a minimum of pressure is put on the headstock. The other end of the blank is punched at the midpoint with a small metal punch.

2. The block for the head of the gavel is usually 2¾″ (11/4's) square and 5″ long. As with many other spindle turners, Andy dislikes gluing up blocks; so he is always looking for extra-thick stock.

After you have cut the head block to size, you should drill a ⅜″-diameter hole through the middle for holding the handle. The hole is actually for a threader that is used to tap threads into the head. This will be discussed in more detail shortly. If you don't have a threader, you can simply glue the handle into the head (Illus. 105).

3. Now mount the handle blank on the lathe using a four-pronged drive center and a revolving center mounted in the tailstock. Illus. 106 presents the handle design and shows the various cuts you will need to make. In order to minimize confusion during the turning process, you may want to look over Illus. 106 and the various procedures that follow before you begin to turn.

After rough-turning the blank with a 1¼″ roughing-out gouge, Andy uses a parting tool to cut the two shoulders at points 1 A and 1 B (in Illus. 106). These shoulders should have an approximate diameter of ⅝″. You will find that a vernier caliper can be very useful with this type of project.

The distance between points 1 A and 1 B is also about ⅝″. The distance from point 1 A to the edge of the spindle at the live center is about 6½″. You should remove the waste between points 1 A and 1 B with a parting tool to

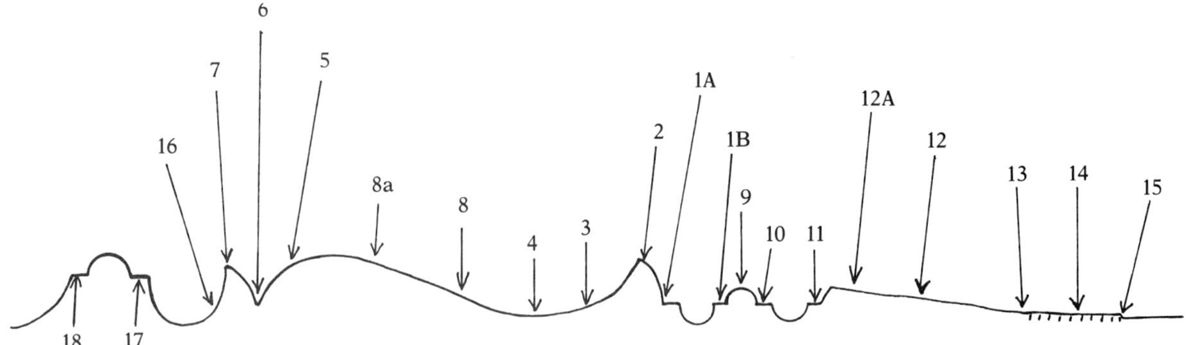

Illus.106. Gavel-handle schematic

allow the placement of a support or steady rest. While crude, Andy's rest does work (Illus. 107). He applies a little beeswax to the contact point to avoid any burning of the spindle. Eventually, this area between 1 A and 1 B will have a small cove cut into it, thus eliminating any charred wood that may develop.

4. With a ⅜″ fingernail gouge, Andy cuts a ½″ wide bead at point 2. He accomplishes this by making thin slices, starting at the top of point 2 and going down to point 1 A. When the gouge ends up at point 1 A, it's on its side.

5. Andy then cuts a deep cove at point 3. He starts at the apex between points 2 and 3 and makes thin cuts, going from right to left. You should start out using the gouge on its side and make scooping cuts. Light cuts are also made from point 4, going from left to right. These cuts also go downhill and are made with a scooping action.

6. The next area to be turned is point 5, which is rounded over with the ⅜″ fingernail gouge. Point 6, the inevitable "tight" area, is approximately 4″ from point 1 A. Again, using the ⅜″ gouge, Andy starts rounding over at point 7. He rounds down into the tight area at point 6. The final cleanup at point 6 is done with a 1″ skew chisel. So that you don't have to change tools, you can do this final cleanup later.

7. The next part of the spindle you need to work on is the point-8 area. You can do this with a roughing-out gouge, a ⅜″ fingernail gouge, or a ½″ skew chisel. If you use a roughing-out gouge, you may need to go over the area with abrasive paper. However, Andy doesn't like to use abrasive paper unless it's absolutely necessary.

8. The next procedure is to cut the bead at point 9. Before he cuts the bead, Andy forms the shoulders at points 10 and 11. These shoulders have an approximate diameter of ½″. The bead at point 9 is about 5⁄16″ wide and is cut with a ¼″ fingernail gouge. Andy uses the ¼″ gouge much as one would use a skew chisel. He starts at the top of the bead and cuts to the left with a flick of his wrist. He uses the same procedure when going to the right. The gouge will be lying on its side when he gets to points 1 B and 10. As you make the cuts, it's imperative, of course, that you ride the bevel. After you've made the shoulders, cut the cove at point 10.

Illus.107. Homemade gavel-handle spindle support

9. You can make the taper at point 12 with the roughing-out gouge or pare it down with a skew chisel. Before cutting the taper from point 12 A, you need to determine the diameter between points 13 and 14. If you plan to use a threader, the diameter needs to be ½″. You will want to use a vernier caliper for this measurement. The length between points 13 and 14 should be at least 1¼″. If you plan on gluing the handle into the head, a ⅜″ diameter will be necessary, assuming, of course, that you've drilled a ⅜″-diameter hole into the head block. You can cut the area using either a roughing-out gouge or a 1″ skew chisel.

10. The small cone taper starting at point 15 should have a diameter of less than ½″. If it isn't smaller in diameter than the area between points 13 and 14, the threader will mark its surface. Use a 1″ skew chisel to plane the taper.

11. With the tailstock area of the spindle turned, Andy begins cutting a cove at point 16 that is approximately ⅜″ in diameter. When you make this cut, use a ¼″ fingernail gouge and be especially careful.

Cut the shoulders at points 17 and 18 to a diameter of about ½″. Andy usually makes the bead between these points approximately ½″ wide and from ¾″ to ⅞″ in diameter.

12. While you may need to do some finishing cuts near the drive center, the spindle is now ready for any sanding that is required (Illus. 108). I have seldom seen Andy use abrasive paper on his spindles; when sanding is necessary, he uses a sanding screen that is used in auto body shops. Called Wetordry Fabricut, this sanding screen is manufactured by 3M and is available in a range of grits. I often use this abrasive screen myself.

13. After sanding the spindle, you should work the ends of it down to a very narrow diameter with a gouge or skew. Then remove the spindle and cut off the nubs on the band saw. Be careful when you are cutting on the band saw that the blade doesn't grab the round spindle from your hands. Lightly sand the ends to remove any marks from the saw blade.

14. The last step before finishing is to cut the threads into the end of the handle. Andy uses a standard ½″ wood threader for this procedure (Illus. 105). After you have started the threads, remove the guide block from the threader so that you can cut the threads to point 13. Rather than using oil on the threader, Andy prefers beeswax. He finds it more effective and it doesn't create the mess caused by oil.

15. Now you need to turn the gavel head. Mount the block between centers and, using a 1¼″ roughing-out gouge, turn the square blank round. Illus. 109 shows the various cuts that Andy makes when he turns the gavel head. You

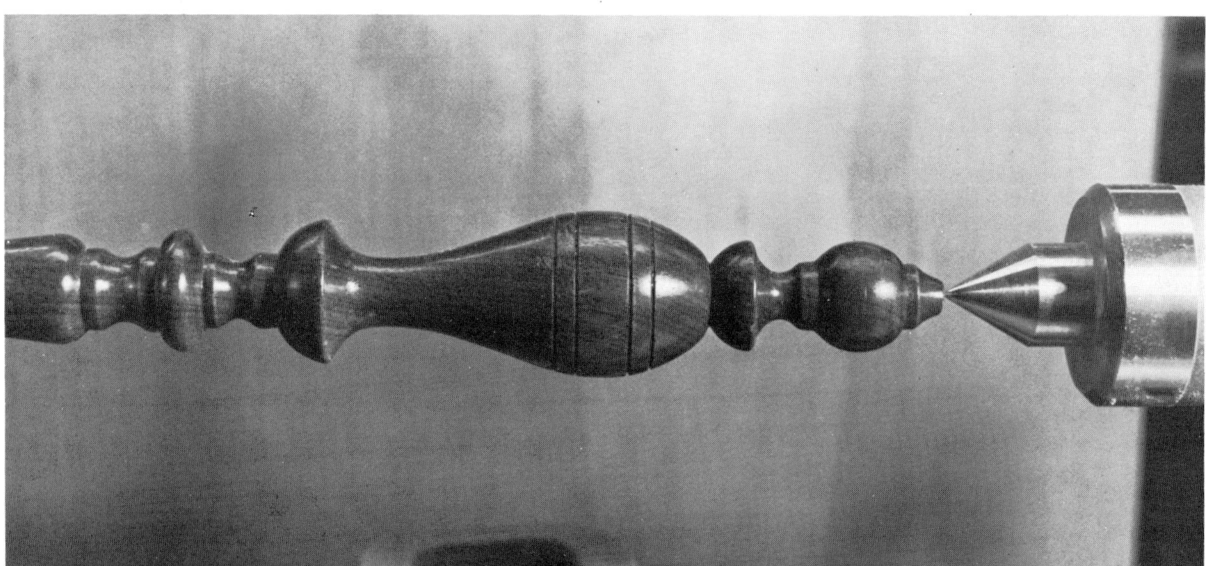

Illus.108. Turned gavel handle

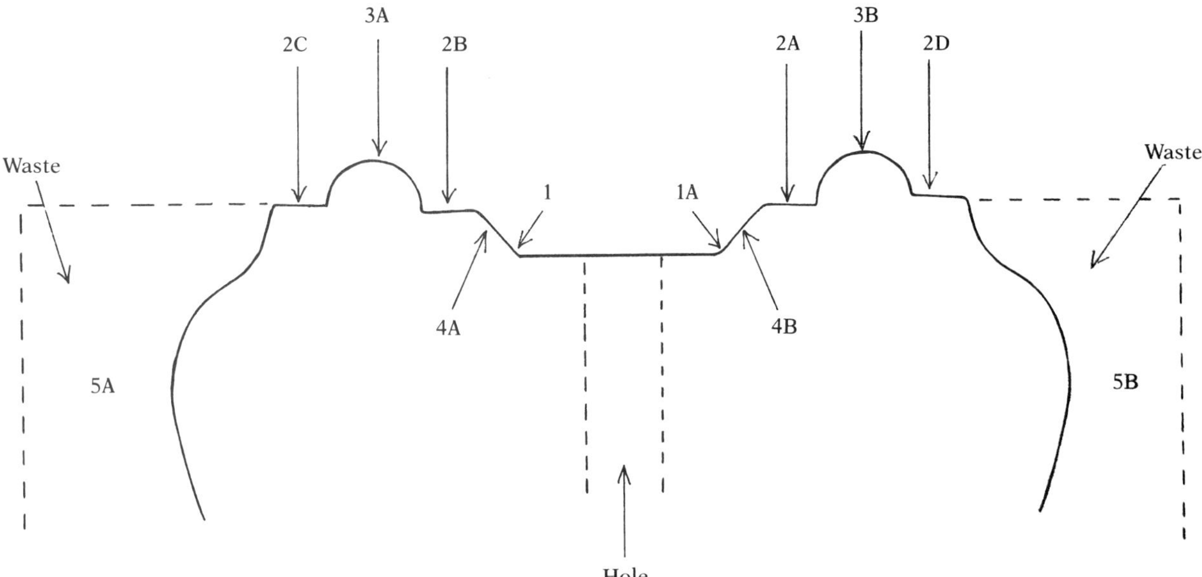

Illus.109. Gavel head schematic

should use this illustration as a reference guide for the subsequent tasks.

16. Use a parting tool to make two cuts at points 1 and 1 A. Cut the area between these two points down to a 1½″ to 1¾″ diameter. To check the diameter, you will want to use a vernier caliper. You can vary the diameter, depending on whether you want a thick, sturdy head or a thinner, lighter head on the gavel. The diameter at points 1 and 1 A should, of course, be the same.

The predrilled hole in the area between points 1 and 1 A will not catch a tool. You can easily turn over the hole. There should be an equal distance in this area on each side of the hole—usually a distance of between ¾″ to 1″. You can tap the threads into the gavel head at this point or wait until you have completed the turning. Remember to countersink the hole at both ends when you tap the threads. This will prevent your tap from splintering the wood. You may want to review the instructions that came with your threader.

17. Using a parting tool, cut the shoulders at points 2 A, 2 B, 2 C, and 2 D. You can make the diameters of these shoulders all the same. Or you can vary the diameters, depending on whether you want a large or small drop-off at points 1 and 1 A. You can also vary the diameters so that you will have tall beads at points 3 A and 3 B. Andy leaves a ⅞″-wide area between points 2 C and 2 B and points 2 A and 2 D. With this width, you can turn a fairly large-size bead in these areas. Andy also initially makes the shoulders wider at points 2 A and 2 B so that he an make an angled cut at points 4 A and 4 B.

You should remove the waste material from point 2 D to the tailstock and from point 2 C to the headstock. For this procedure, you can use either a parting tool or a roughing-out gouge placed on its side.

18. The next task is to cut the beads at points 3 A and 3 B. As with most spindle turners, Andy takes pride in his ability to cut a perfect bead. He uses a ¼″ fingernail gouge to make the two beads. While a ¼″ gouge may seem rather small for making large beads, Andy prefers it. You should use the gouge much as you would use a skew. Andy begins by cutting the corners off at points 3 A and 3 B. You should make about three thin cuts at each corner. Follow these cuts by one final cut from the middle of the bead to point 2 A and then from the middle of the bead to point 2 D. Use the same cutting procedure for the other bead.

When making these cuts, remember to ride the bevel and use a lot of wrist action. The gouge should be on its side when you have completed the cut. Using the gouge in this manner will prevent you from tearing the end grain.

As you may know, torn end grain is almost impossible to correct with abrasive paper. The sanding that is required to correct it inevitably results in also changing the shape of the piece. Andy is not adverse to sanding, but he reserves the use of abrasives for side grain.

19. After you have cut the two beads, you will need to make the angled drop-off at points 4 A and 4 B. Here again, the ¼″ gouge is used as a skew. The cuts are made by two or more light passes on each side. Be certain to rub the bevel.

20. Shape the ends of the gavel at points 5 A and 5 B. Andy tends to shape them with a slight contour from the shoulders at points 2 C and 2 D.

To shape the ends, you can use the ¼″ gouge again. Since this is tough end grain, several cuts are required in addition to considerable wrist action. You also should be using a sharp tool. On the final cuts, Andy rides the bevel down, with the gouge on its side, to the point of almost cutting the spindle from the lathe. He leaves a diameter near the nub, on both ends, of about ¼″ (Illus. 110).

21. Remove the spindle from the lathe and band-saw the nubs from each end. Cut as near as possible to the ends of the gavel head. Then clean up the cuts with abrasive paper.

22. It's best to turn the gavel base from 1″ (4/4's)-thick stock that is at least 4½″ in diameter. You

Illus.110. Turned gavel head

can use a screw chuck, double-face tape, or white glue and turn the assembly on a face-plate. Andy likes to dress up the base by using his homemade "cheater," which is described in Chapter 3 and shown in Illus. 24.

Now you may want to refer to Chapter 43 for a discussion of finishing products and procedures.

32
ASHTRAYS

An ashtray is a project that is both decorative and functional. While smoking is clearly falling into disrepute, ashtrays still remain in high demand in many households. There are endless designs that can be developed for ashtrays, but it isn't always so easy to find glass inserts that are the right size. The way I deal with this problem is by purchasing regular glass ashtrays that are 3¾" in diameter and ⅞" thick from local discount dealers. I've found that this size, especially the thickness, is ideal for a wooden ashtray made from 1" (4/4's)-thick stock. You can also use a glass ashtray this size when turning ashtrays from thicker stock. You may want to refer to Illus. 111 to note some of these options.

While this ashtray is specifically designed for cigarettes, if you increase the size of the base and its insert you can use it for cigars. If you want to make an ashtray for a pipe, you can

Illus.111. Ashtrays

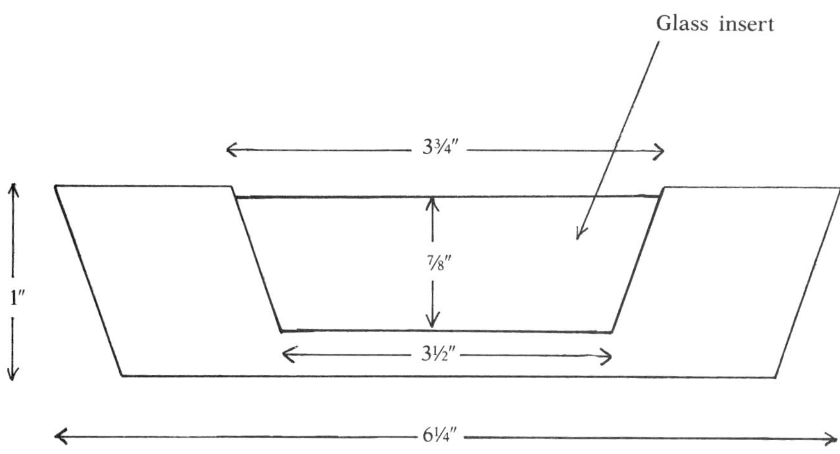

Illus.112. Ashtray and dimensions

obtain small cork knockers to glue into the glass insert. There is a wide variety of glass ashtrays on the market that are suitable for cigars or pipes. However, if you use larger ashtray inserts, you will need to increase the thickness of the wood accordingly.

To assist you in turning this project, refer to Illus. 112, which shows a drawing of an ashtray along with some approximate dimensions.

TASKS:

1. Pattern and cut a 1″ (4/4's)-thick block that is at least 6½″ in diameter. If you would prefer having more wood surface surrounding the glass insert at the top, increase the diameter of the block. For example, I try to have at least a 1″-wide area of wood that surrounds the insert. This is the only area on the entire ashtray where you can showcase the wood.

2. Depending on your own preference, secure the block to a 3½″ pine or plywood disc using either double-face tape or white glue. Attach a 3″ faceplate to the assembly. You may want to refer to Chapter 1 for a discussion of these procedures. Then mount the assembly on the lathe.

3. True the block and then shape it using a 1″ roundnose scraper or another tool of your choice. You may want to refer back to Illus. 112 to note how the side of the block is tapered inward at the bottom, although you may prefer to make the side concave or some other shape. Now slightly roll the top edge in preparation for turning the top surface.

4. Align the tool rest in front of the block. Before making the cut for the glass-insert area, check the diameter of the insert with a vernier caliper. You want the insert to fit neatly into the recessed area without sliding around. You should also check the diameter at the bottom of the glass insert. These two measurements will allow you to have a tapered recess, where the insert will neatly rest.

Using a ½″ square-nose scraper, begin removing wood from the middle area of the block. Do not exceed the thickness of the glass insert. The bottom thickness of the ashtray should be approximately ⅛″. As you remove wood from the area, monitor the diameter with the caliper. As you make the final cuts, be certain the inside wall tapers to the correct diameter at the bottom. Then stop the lathe and place the glass insert into the recess to check for fit. The top surface of the block and the top edge of the glass insert should be flush.

5. If necessary, sand the various surfaces using a series of abrasive grits. You may want to slightly roll the top-inside edge that butts against the glass insert. This will make removing the glass insert, when it's full of ashes, much easier. I also go over all the surfaces with (0000) steel wool.

6. Remove the assembly from the lathe. If you used tape to secure the block to the disc, remove it very carefully with a chisel; if you used glue, remove it with a band saw. Sand the bottom surface to finishing readiness. Then turn to Chapter 43 for a discussion of finishing products and procedures.

33
SOAP DISH

This small project emerged from necessity. Since most modern bathroom sinks do not have any kind of indentation for holding soap, you either have to place the soap on the top surface of the sink, where it leaves a mess, or buy some makeshift soap dish. I solved the problem in our bathrooms by turning small soap dishes from beech. While you can use any type of wood for this project, beech tends to have the most durability when used around water.

A critical issue in turning soap dishes is making sure they are protected with an appropriate finish. You need to use a sufficient amount of a type of finish that will resist the range of chemicals that are in various soaps. As you will discover in Chapter 43, I tend to use Deft, a quick-drying lacquer, on most of my turnings, including soap dishes. To maintain the finish over a period of time, it's a good practice to remove the soap residue from the dish on a regular basis. If the finish is not maintained, the chemicals in the soap will penetrate it, eventually causing the wood to turn black.

I generally use either cutoffs of beech or a lower grade of beech for soap dishes. When cutting boards to length, lumber dealers frequently will give away the remaining cutoffs or sell them quite cheaply. As a number of later projects will also indicate, this can be an excellent source of material.

You should pattern a 1″ (4/4's)-thick piece of stock to a diameter of approximately 5″. Secure it to a 3½″ pine disc and a faceplate using either double-face tape or white glue. Then mount the assembly and turn it to the shape you desire. However, it's best to leave the bottom of the dish rounded so that you can easily remove a bar of soap. For specific procedures, refer back to Chapter 1.

Illus.113. Soap dish

34
SQUARE OR RECTANGULAR GRIMPLE HOLDERS

An enjoyable project to make, especially from scraps, is what I call grimple holders. In case you are unfamiliar with the term, "grimple" refers to any small item that needs to be picked up, put someplace, and saved. Buttons, pins, paper clips, and pennies are all examples of grimple. Since grimple tends to mount up, appropriately placed grimple holders are needed in every household.

As you can tell from Illus. 114, grimple hold-

Illus.114. Grimple holders

ers can be turned from different types of wood. I've found that scrap ends of boards or cutoffs from lumber dealers are ideally suited for this project. This type of material usually has some broken edges, knots, or bark—or something else that will give the piece some character. There are no required dimensions for grimple holders. They can be made from stock of any thickness, in squares or rectangles. Grimple holders are ideal for making use of the wood on the bottom of your scrap box.

Since the wood used for grimple holders will often have sharp edges, you would be well advised to use a glue disc and white glue to secure the block to a faceplate. However, with some blocks that are small and not too thick, double-face tape may be most appropriate. Mount the assembly on the lathe and then turn out a rounded area to a diameter that takes up most of the block's surface. Use a 1″ or ½″ roundnose scraper. Leave the bottom of the turned-out area round so that the grimple can be easily removed. After removing the turned block from the glue disc, sand all the surfaces using a belt or pad sander. Sand the turned area by hand while the piece is either on or off the lathe. For finishing instructions, refer to Chapter 43.

35
PAPER-CLIP DISHES

With this small desk-top project, you can choose from an array of possible designs. And you can easily turn the project from scrap material. Even though the dish is intended for paper clips, obviously you can use it for pins and buttons and anything else that might fit. I originally made these small dishes to serve as companion pieces to the round pen bases presented in Chapter 37. Especially if you match the wood, a paper-clip dish and a pen secured in a round base can be an excellent gift.

The dishes in Illus. 115 has been turned from scraps of wood that are ½″ to 1½″ thick. You should determine the diameters primarily in relation to the size of your available scrap material. Usually I will turn the dishes to a 3″ to 4″

Illus.115. Paper-clip dishes

diameter. Since a desk-top paper-clip dish doesn't need to be very large, these suggested dimensions should be more than adequate.

As a rule, I turn these small dishes using double-face tape. Because the blocks are small, double-face tape is ideal. However, to minimize cost or to make sure the block is secure, you may want to use glue discs and white glue instead. Either way is effective for faceplate-turning this project.

Some of the designs that I turn are planned, but most simply emerge during the turning process. Since you have a minimum of investment in materials, you can turn this project as your whim moves you. If you are planning on turning a dish-and-pen set, you may want to make the two pieces have similar shapes. However, I've found that two pieces can look compatible even if they have very different shapes. In any event, have fun turning these small dishes and try some unusual designs.

I generally turn the inside bottom round to facilitate removing paper clips or anything else that might be placed in the dish. It's best to turn these small dishes using a ½" roundnose scraper. If you want to have a flat bottom on some designs, you should use a ½" square-nose scraper.

For information on finishing products and procedures, refer to Chapter 43.

36
DESK-TOP PEN-HOLDER BASE WITH PAPER-CLIP CONTAINER

This rectangular base accommodates a pen and its funnel and also has a turned area for paper clips. To turn the paper-clip indentation, the block must be mounted and turned off-center. The pen and funnel that I use for this project are relatively inexpensive and available from

Illus.116. Pen and base with paper-clip holder

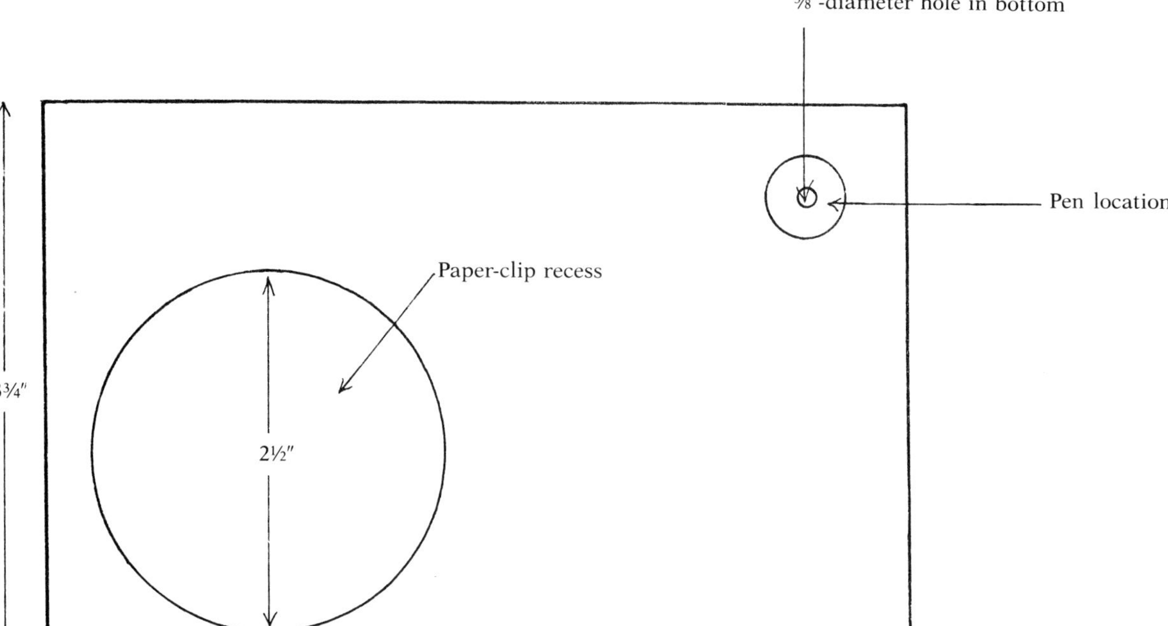

Illus.117. Pen base

many mail-order suppliers. You can also obtain this type of pen and its holder in gold and other metals. If you want to add a special touch for a gift, you can have a small brass presentation plate engraved and then secured to the top surface.

I generally try to find an unusual piece of scrap material for the base. For example, the base in Illus. 116 is made from spalted honey locust that had been infested with insects. Most commercial pen sets that I see in stores are made from a perfect piece of wood; the uniformity in the wood almost makes them look as if they're made of plastic. But I prefer to use an unusual piece of scrap material that is laced with natural aberrations. Bases made from this kind of material are not only more attractive, but they also stand out on a desk and can be great conversation pieces. You can find this kind of material in the lower grades of hardwood.

Although you can use any dimensions for this project, you may want to follow the suggested dimensions in Illus. 117. According to the drawing, you will need to drill a ⅜″-diameter hole into the bottom of the base to accommodate the screw head that holds the pen funnel in place. More about this shortly.

TASKS:

1. Prepare a rectangular block from 1″ (4/4's)-thick stock that is 5″ long and 3¾″ wide. You may want to increase the overall size of the project according to your own design or your available wood. I've found these dimensions to be about the minimum required for this type of project.

2. Since I turn the paper-clip area near one side, the block needs to be mounted off-center on a glue disc. Prepare a 3½″ glue disc from either pine or plywood. Whenever I turn a block off-center, I use white glue rather than double-face tape. I suspect that the tape would hold, but I prefer using white glue and a glue disc for safety precautions. After spreading glue on the disc, place the disc directly on the bottom of the block, beneath the area where you will be turning the recess for the paper clips. The glue

disc may extend slightly beyond the edge of the block. Now clamp the assembly and allow the glue to dry. Then secure a faceplate to the disc and mount the unit on the lathe.

3. When turning off-center, I reduce the lathe speed to 990 rpm. On occasion, I would like to turn at an even slower speed, depending on the size of the block; but unfortunately, this is as slow as my pulleys will permit. *Be very careful to keep your hands out of the way of the rotating block*. Also, be certain the block clears your tool-rest base before you turn on the lathe.

Using a ½" roundnose scraper, turn the paper-clip area to the suggested (or preferred) diameter. I generally make the area at least ¾" deep. After you have turned the area, you can sand it while the block is still on the lathe or, if you prefer, after you've removed it.

4. Cut the base from the glue block using a band saw. If you saw slightly into the bottom surface of the base, all of the glue will be removed. If you don't have a band saw, chip the glue disc off using a wood chisel; then clean up the remaining glue using a large belt sander and a coarse abrasive belt.

5. The pen funnel is usually secured to the base with a screw. Determine where you want the pen to be located on the base and then drill a ⅜"-diameter hole into the bottom surface, directly under that spot. The drilled hole only needs to be deep enough to accommodate the screw head. Now determine the diameter of the funnel screw and then drill another hole through the middle of the ⅜" hole. This second hole will allow the screw to penetrate the base and go into the funnel. If the funnel screw is too long for the thickness of the base, saw a portion of it off using a hacksaw. Do not secure the pen funnel in place until you've completed the project.

6. Prepare all surfaces for finishing by using a range of abrasive grits. For some suggestions regarding finishing products and procedures, refer to Chapter 43. Once the base has been finished, secure and align the pen funnel on it.

37
ROUND BASE FOR PEN HOLDER

This desk-top project can be used alone or as a companion piece to the paper-clip dish presented in Chapter 35. Although there are all kinds of commercial pens available, I generally buy the plastic units that come with funnels. You can also use this type of base for a pen you've turned yourself. (For instructions on turning ball-point pens, refer to Chapter 2.)

You can easily make this pen base from scrap material. The bases in Illus. 118 are made from scrap pieces, which vary in thickness from ¾″ to 2¼″ and in diameter from 2¾″ to 3¼″. As with many turned projects that I make, the design and dimensions of this project are dictated by the available scrap material.

TASKS:

1. Pattern a block that is at least ¾″ (3/4's) thick and has a diameter of approximately 3″. Since it's best to turn the block on a screw chuck, drill

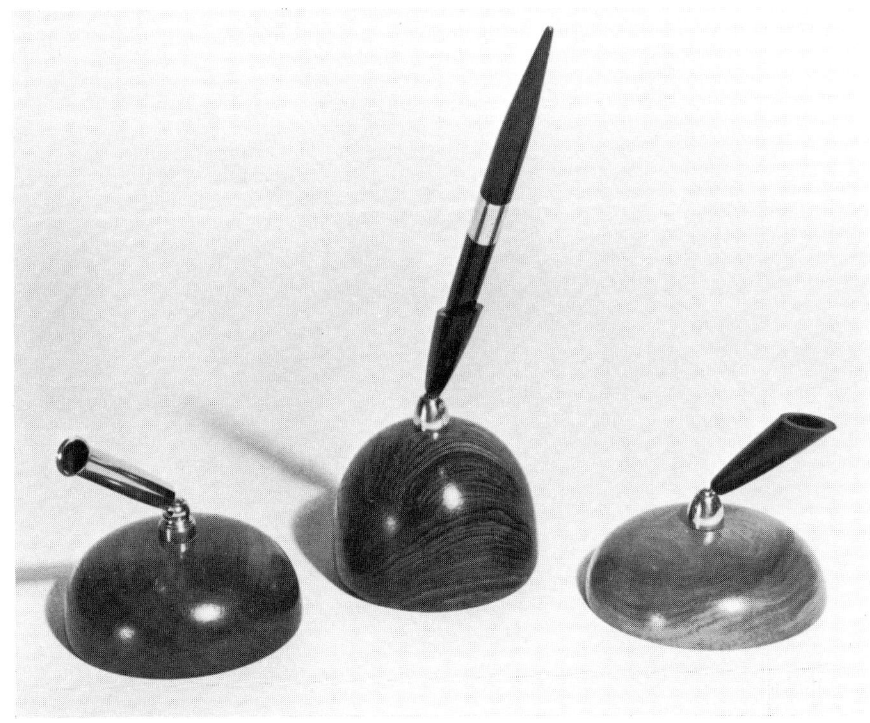

Illus.118. Round bases with pens

a hole through the middle of the block that will tightly accommodate the screw. If you don't have a screw chuck, you may want to review the instructions for making one that are included in Chapter 1. This same drilled hole in the middle of the block will be used later to accommodate the funnel screw.

2. Attach the screw chuck to the headstock and then secure the block to the chuck. Using a ½″ or 1″ roundnose scraper, true and shape the block. You may want to refer to Illus. 118 to note the rounded shape of the block. I tend to turn a rather plain rounded base, but you may prefer something a bit more decorative.

Once you've turned the shape to your satisfaction, sand the surface to finishing readiness.

3. With the turned base removed from the chuck, you need to drill a hole into the bottom to accommodate the head of the funnel screw. Usually a ⅜″-diameter hole is more than adequate. The depth of the hole is determined by the actual length of the funnel screw. When the funnel screw is in place, it should extend beyond the top surface of the base far enough for the funnel to be screwed tightly in place. If you drill the ⅜″-diameter hole too deep, place small washers over the funnel screw or saw off a small piece of its length. Remember, you are drilling this hole into the lag-screw hole; so take your time. A drill press is almost mandatory for this procedure. Do not secure the pen funnel to the base until you've completed the finishing. For information on finishing products and procedures, refer to Chapter 43.

38
SOLID APPLE

Although you can use it as a decorative piece, this solid wood apple makes a great paperweight. With this project, you will get the chance to use a number of tools in addition to the lathe. After you've turned the apple on the lathe, you will shape it with a large belt sander. Then you will cut and shape a stem and a leaf using small sanding cylinders and a Moto-Tool or Flex-Shafts tool.

For this project, you can use a log from a woodpile. I made the apple in Illus. 119 from a black walnut log that was about 5″ in diameter, which I found in a friend's woodpile. Because of the diameter, I turned the apple with the grain

Illus.119. Solid walnut apple

running parallel to the lathe bed. This gave me a sufficient diameter and the length of the log provided adequate height. I made the stem from a scrap piece of Indian padouk and the leaf from a green streak in a scrap of poplar.

TASKS:

1. Prepare a block that has a 4″ to 5″ diameter and is approximately 4″ long. I suggest you try to find a piece of a log for the project. I was fortunate in coming across a black walnut log, but any local wood should look just fine. Another option, if you can afford it, is to buy a piece of Indian padouk cut to these dimensions; then you can turn a red apple.

2. Secure a 3″ faceplate to the end of the block using wood screws that are at least ¾″ long. Since the wood screws are going into end grain, they should penetrate quite far so that the block will not pull off during turning. Now mount the assembly on the lathe and bring the tailstock with a revolving center in place forward into the block. This setup will provide security for the block during the turning process.

3. Using a 1″ roundnose scraper or a roughing-out gouge, true the block. You may want to examine a real apple to note its shape. Now rough-shape the apple on the side-and-bottom edge; then begin a slight roll on the top edge of the block.

Remove the tailstock and align the tool rest in front of the block. Turn the top surface slightly round and then make a recess where you will place the stem. I usually make this recess at least ½″ below the top surface. Use light pressure with your tool so that you won't force one of the faceplate screws loose. It's not necessary to sand the surfaces at this time.

4. Remove the assembly from the lathe and the faceplate from the block. The next task involves shaping the apple using a 6″ × 48″ belt sander (or a small belt sander) and a 60-grit abrasive belt. *It's a good practice to wear leather gloves during this procedure.*

This is another time when it would be a good idea to note the shape of a real apple. I usually use delicious apples as my models. Note the various ridges along the sides and the knobs on the bottom. Before you begin to shape your apple, remember that real apples are not perfectly symmetrical.

Using the roller end of the belt sander, shape the apple to its final form. Be careful that the belt doesn't pull the apple from your hands. Don't worry about the wood grain or the surface as you're shaping because you will be sanding the surfaces with a small electric pad sander in the next task. Be sure to shape the top surface of the apple in addition to the sides and bottom. After you've shaped the bottom surface, set the apple on a flat surface to make sure it stands erect; if it doesn't, remove wood from the bottom surface until it does.

5. After you have properly shaped the apple, bring all surfaces to finishing readiness with the electric pad sander. I usually begin with a 100-grit abrasive. Sanding with the grain, remove all the scratches made by the coarse abrasive belt. Follow the 100-grit abrasive with 150-grit and then finish with 220-grit. You may also want to rub the surface down with (0000) steel wool.

6. Using a ¼″-diameter high-speed drill bit, drill a hole at least ½″ deep in the top center of the apple. This is the hole where you will place the stem.

7. As I've indicated, the black walnut apple in Illus. 119 has an Indian padouk stem. But you may prefer using another wood, depending on the wood you used for the apple. I made the leaf from a small piece of poplar. Usually poplar has green streaks running through it. I cut out a small section of the green to use for the leaf. You may want to refer to Illus. 120 to note how you should rough-cut the stem and leaf.

Pattern the leaf on the wood's surface. It should be approximately 1¼″ long and 3⁄16″ thick. Now cut out the leaf using a scroll saw or another saw that's similar. Drill a 1⁄16″-diameter hole into one end of the leaf. The hole should penetrate at least ½″.

Pattern and cut the stem from an appropriate wood to the size and shape indicated in Illus. 120. The stem should be at least 5⁄16″ thick. Now drill a 1⁄16″ hole into one of the branches of the stem. It should penetrate the branch by at least ½″. This hole will eventually receive the wire attached to the leaf (Illus. 121).

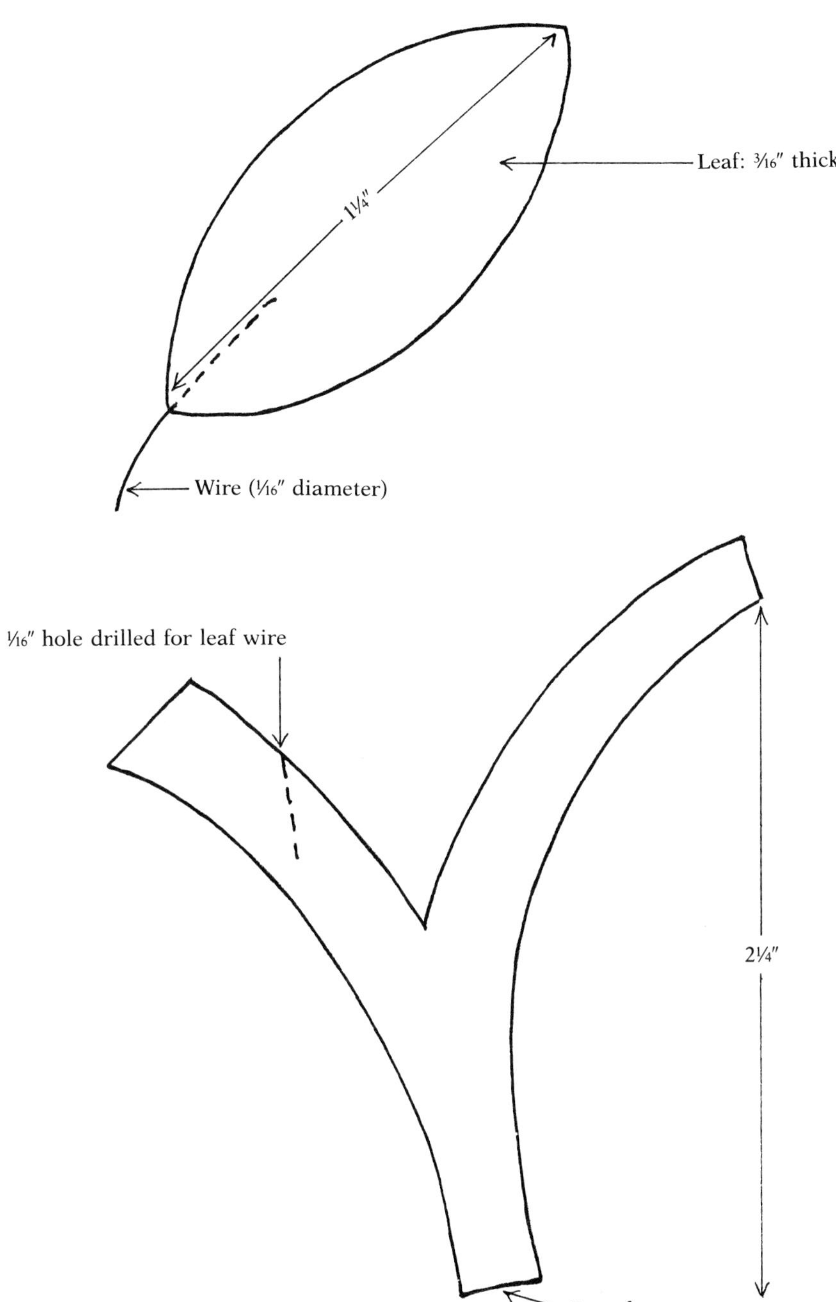

Illus.120. Stem and leaf for apple

To shape the leaf and the stem, I use a Flex-Shafts tool or a small Moto-Tool and sanding cylinders secured to a mandrel. Shape the leaf so that it has sharp edges and a thick center. Do not remove so much wood that you expose the drilled hole. Vary the shape of the stem so that some portions are rounded and others are somewhat square. As you shape the stem, also vary its diameter. Make it resemble a real stem as much as possible. Taper the bottom section of the stem so that the bottom ½″ will fit tightly into the hole drilled in the apple; then test for fit. Handle the stem carefully because it becomes quite fragile after it has been shaped.

8. Cut a piece of 1⁄16″ wire to a length of approximately ⅝″. Place white glue on one end of the wire and then insert that end into the hole in

the leaf. You should also place a drop of glue on the other end and then insert that end into the hole in the branch. Adjust the leaf on the branch so that it looks natural.

Spread some wood glue on the tapered end of the branch and then insert it into the drilled hole in the apple. Wipe away any excess glue that may have been squeezed out onto the top surface of the apple. When the glue is dry, you should be able to pick up the apple by the stem. Now refer to Chapter 43 for some finishing ideas.

Illus.121. Block, stem, leaf, and wire

39
SMALL CANDLEHOLDERS

If you enjoy spindle turning, you will probably want to make these small candleholders. I'm especially fond of turning these small pieces from pine. They can, of course, be turned from any wood. I began turning them from pine when I picked up a box of 2″ × 4″ scraps at a construction site. It's amazing how much wood is discarded at construction sites that can be put to good use on the lathe.

The candleholder is designed for a candle with a 15⁄32″-diameter base. A candle this size is usually about 4″ long. While you can use other sizes, this diameter and length is in proportion to the candleholder.

Illus.122. Pine candleholders

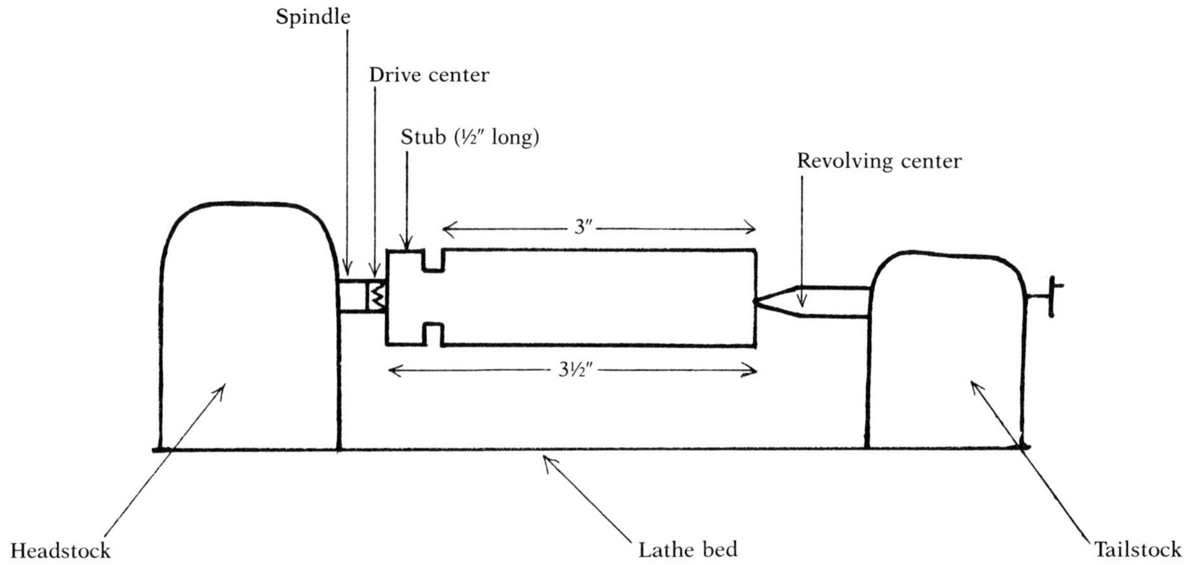

Illus.123. Candleholder blank between centers

I turn these small candleholders between centers using a standard four-pronged drive center and a revolving center in the tailstock. You may want to refer to Illus. 123 to note the lathe setup as well as some approximate dimensions.

TASKS:

1. Using 2″ × 4″ scrap material, cut chunks that are at least 3½″ in length. If you want to have a supporting stub on the tailstock end in addition to the headstock, cut 4″-long chunks. Then rip the chunks into 1½″ squares. It's a good idea to prepare a quantity of blanks because you probably will want to make a few candleholders once you begin turning. You should make diagonal lines on both ends of the blanks for proper centering of the drive center and the revolving center.

To speed up the turning process, I make indentations on all the blanks with the drive center. With one end of the blank placed on a concrete floor, align the drive center on the diagonal lines; then center it and tap it with a wooden mallet. Now secure one of the blanks between centers.

2. With a 1¼″ roughing-out gouge, turn the blank round. You should be careful with the tools around the four-pronged drive center. If necessary, leave the stub square. Using a parting tool, make a separation cut between the base of the candleholder and the front edge of the headstock stub. I usually make the stub about ½″ long. If you cut the blank for a stub at the tailstock end, repeat the procedure.

I do not always use a stub at the tailstock end of the blank. The small indentation made by the revolving center marks the entry point for drilling the hole for the candle. This also assures you of an accurately placed hole. If you use too much tool pressure during the turning process, the revolving center will begin to flop around on the soft pine's surface. Even though the point enters the end grain of the blank, too much tool pressure will widen the indentation and this, in turn, can throw the spindle off-center. You may need to increase the pressure from the tailstock and support the spindle with your hand while turning. Don't use too much tailstock pressure or you will break the spindle.

When you are turning the candleholder, try different gouges. I primarily use ¼″- and ⅜″-deep fingernail gouges. On occasion, I will use a ½″ fingernail gouge. Pine is a good turning wood and it's fun to use for experimenting with various tools and cuts. You want to be certain that your tools are sharp and that you ride the bevel. Pine is rather coarse and can easily tear out. In addition to the gouges, you might want to experiment with a ½" skew chisel. This is an excellent project for refining your skill in cutting beads with a chisel. You may want to refer back to Illus. 122 to note the shapes of the candleholders. Also, Illus. 124 shows a turned candleholder between centers.

3. Depending on the condition of the surface, some sanding may be required. To avoid scratching the pine, use 150-grit or finer abrasive paper. I also go over the surface with (0000) steel wool, which tends to polish the surface and bring out the character of the wood.

4. Using a parting tool, separate the candleholder base from the stub. Be sure to hold your hand under the candleholder so that it doesn't drop on the lathe bed. Pine dents and scratches very easily.

5. Using a 15⁄32"-diameter high-speed steel bit, drill the hole for the candle into the top surface. The revolving-center indentation should be used as the entry point for the drill bit. You should use a drill press for this procedure. If you're using a candle that is a different size from the one suggested, check its base diameter and then use an appropriate bit for the hole.

6. Finish-sand the candleholder's top and bottom surfaces. Since these areas are end grain, they may require a little extra work to bring them to finishing readiness. Then refer to Chapter 43 for finishing procedures.

Illus.124. Candleholder blank being turned

40 KEY HOLDERS WITH CHAINS

Key holders are not only fun to make, but they also give you the opportunity to further develop your turning skills. You can learn a great deal about the various gouges and chisels with this kind of project. And you can also practise turning beads and coves without having to worry about wasting wood, because you can use scrap material for this project. When you have completed the project, you can distribute the key holders among your friends.

To turn these key holders, it's best to use a ½"-diameter Jacob's chuck mounted on a Morse taper. In addition to this project, the next two projects in this book also involve using a Jacob's chuck. If you don't have a Jacob's chuck, you may want to buy one for spindle-turning

Illus.125. Key holders with chains

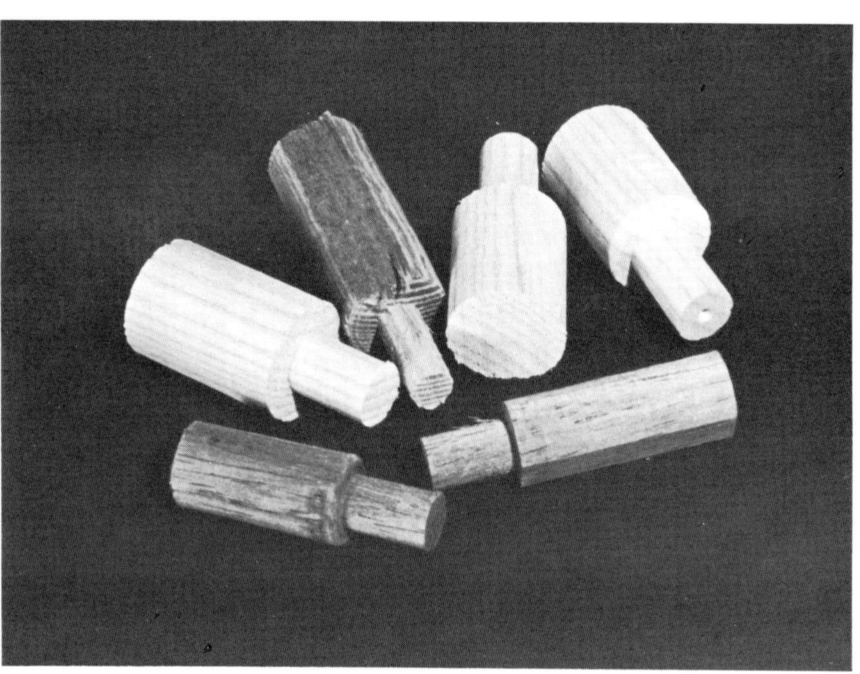

Illus.126. Key-holder spindles with ½"-diameter tenons

small projects. Although you can do the same thing and more with one of the commercial chucks, a Jacob's chuck on a Morse taper is much less expensive.

In order to use the Jacob's chuck, you need to turn a ½"-diameter tenon on the spindle. The tenon should be at least 1" long because, when the chuck is tightened, it tends to force a portion of the tenon from the chuck. Illus. 126 shows a collection of blanks and spindles with ½"-diameter tenons. While turning tenons on each spindle can be a bit of a nuisance, the end result is worth the effort. If you prefer small key holders, you can use ½"-diameter dowels. This, of course, eliminates the need for turning tenons to fit into the Jacob's chuck. Another option is to turn the key holders using a minidrive center and a revolving center. But I'm inclined to use the Jacob's chuck, probably more out of habit than anything else.

A quick method for preparing spindles with ½"-diameter tenons is to turn a blank that is at least 1¼" square and 12" long. Before beginning this procedure, you need to decide how long you want the individual key holders to be when they're finished. I turn key holders anywhere from 2" to 4" in length, depending on the type of wood I'm using. The diameter is another consideration when preparing the spindles. Key-holder diameters can vary from ½" to 1", depending on the design and the length of the spindle.

Cut a blank that is 12" long and 1¼" square. Mount it between centers using a four-pronged drive center and a revolving center. Turn the blank round using a 1¼" roughing-out gouge. Then shut off the lathe and, with a pencil and ruler, mark off the individual spindles and their tenons. Remember that the tenons should be 1" long. Using a parting tool and a vernier caliper, reduce the diameters of the tenons to ½" (Illus. 127). To save wood, I separate the individual spindles and their tenons with a band saw instead of parting them on the lathe.

To hold the keys, I use a standard bead chain. And I use a 9/64"-diameter high-speed steel bit for drilling the key-chain hole through the key holder. The shape and design of the key holder determines where you should drill the hole.

TASKS:

1. Secure a spindle with a ½"-diameter 1"-long tenon in a Jacob's chuck held in the headstock on a Morse taper. Be certain the taper is secure in the headstock. If the spindle proper is more than 3" long, you may want to use the tailstock with a revolving center as support.

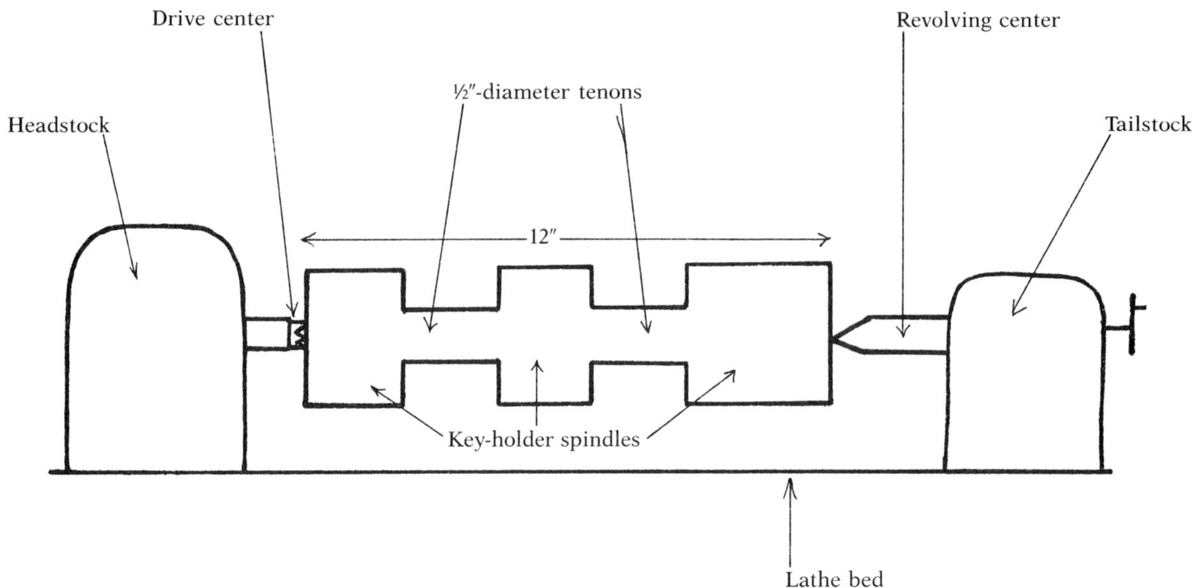

Illus.127. Spindles with turned ½″-diameter tenons

2. Using a 1″ skew chisel, plane the surface of the spindle. Not only will this procedure clean up the surface, but it will also round out the spindle. For this task, you may prefer using a ½″ square or skew chisel; sometimes the smaller chisel is easier to use on short spindles.

I generally turn key holders using a ¼″- or ⅜″-deep fingernail gouge. To maximize the wood's diameter and thus its support, I begin turning the spindle at the tailstock end. If the spindle begins to vibrate, support it with your other hand. I place the tool rest well below the spindle. The tool rest is near enough to the spindle so that I can lay the bevel of the gouge on

Illus.128. Key holder in Jacob's chuck

the wood's surface. I raise the tool rest to slightly below the spindle when I'm using a chisel. You may prefer to raise it higher in order to use the chisel on the top of the spindle. I tend to apply it to the side.

You can make some beads and coves or other kinds of decorations on the key-holder spindle. You might want to taper and cut off a section of the tailstock end using a skew chisel, just for the fun of it. Try different cuts and angles with your tools. Remember to ride the bevel of your tool. Watch how the wood either accepts or rejects the tool when it's placed on the spindle. Have fun and take advantage of this chance to learn more about spindle turning and wood.

3. Now separate the key holder from the tenon. I usually make a partial bead and simply roll a gouge to the left and ride the bevel down until the key holder separates from the tenon. But you may prefer making a concave, partial cove cut instead. Similar cuts are required if you used the tailstock for support. Don't, however, cut all the way through the spindle. You can cut off this end with the band saw. Incidentally, if your tools are sharp, no sanding should be necessary (Illus. 128).

4. Decide which end should have the key-chain hole drilled through it. I usually drill it in the end that has the most wood mass. As indicated, a 9/64"-diameter bit is usually adequate for most bead-chain key holders. After you've drilled the hole, refer to Chapter 43 for finishing procedures.

41
TINY BOXES

For this project, you can use scrap pieces of wood that you normally would sweep up from the floor and discard. The bases and lids in Illus. 129 range from ¼″ to ½″ in diameter; the height of the bases and their lids range from ½″ to 1½″. Surprisingly enough, you can actually use standard-size turning tools to make these tiny boxes.

Because of the size of the project, you need to turn a tenon on the lid. The tenon will penetrate a drilled hole in the base. You should use a vernier caliper for turning these tiny boxes. And it will be necessary to monitor the tenon quite frequently during turning. It's very easy to turn away too much wood very quickly on projects this small.

I turn the boxes using a Jacob's chuck secured in the headstock on a Morse taper. Since the chuck capacity is ½″, the box spindles need to have a tenon of this diameter. The tenon should

Illus.129. Tiny boxes

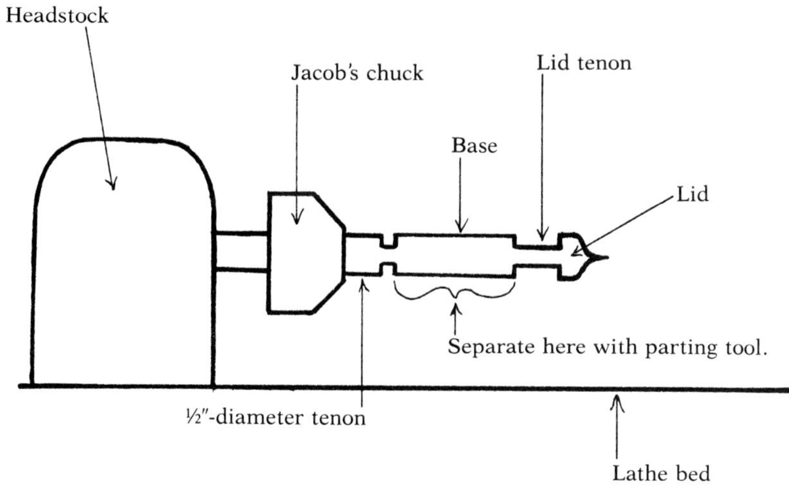

Illus.130. Tiny box in Jacob's chuck

be at least 1″ long to be properly held in the chuck's jaws. For turning the boxes, I run the lathe at 2220 rpm. However, you may prefer to increase or decrease this speed, depending on the type of lathe you use.

You can make the spindles for the boxes from any available scraps, but I often use Osage orange, black walnut, and hard maple. You can prepare the spindles and their tenons following the procedures discussed in Chapter 40. Standard dowels are an ideal source of wood. I frequently use pieces of ½″-diameter dowels that are in the bottom of my scrap box. Since dowels are available in a range of different hardwoods, you can make tiny boxes from many types of wood. I also use dowels that are ⅝″ or ¾″ in diameter and turn a ½″-diameter tenon on them. The tenon can be easily turned by mounting the dowel between centers and reducing a 1″-long section of it to a ½″ diameter. Once you have turned a few of these tiny boxes, you will be amazed at how much wood you have available for this project.

TASKS:

1. For an initial box, cut a 2″-long piece of ½″-diameter dowel. Mount the Jacob's chuck and its Morse taper in the headstock. Slide the dowel spindle into the chuck and secure it in place. Then tighten the chuck jaws all around the chuck so that the pressure on the tenon is even.

2. Align the tool rest to a position for using a 1″ or ½″ skew chisel. Then plane and true the spindle with the chisel. You can also use the chisel to make the top portion of the lid so that you don't have to change tools. However, you may prefer to use a ¼″-deep fingernail gouge to shape the top section of the lid. I usually turn the top section of the lid first.

Examine the spindle and decide how long you want the lid and base. Remember to include the length of the tenon. For example, on a 1″-long spindle, I make the base approximately ½″ long. This leaves a tenon that is about ⅛″ long; then the lid, of course, is ⅜″ long. When I separate the lid and tenon from the base with a parting tool, I cut slightly into the base. Given the thickness of the parting tool, this prevents cutting away too much of the tenon during the separating procedure (Illus. 130).

After you have turned the lid, you will need to cut the lid tenon using a parting tool and a vernier caliper. If the overall diameter of the base and lid is ½″, you should cut the tenon to a ¼″ diameter. This will allow sufficient wall thickness for the base. Also, most shops have a ¼″-diameter high-speed or brad-point drill bit lying around with which to drill the storage area in the base. For this project, the tenon should be approximately ⅛″ long. As you cut the diameter and length of the tenon, monitor the diameter frequently with the caliper. Make very light cuts and remove only a minimum of wood. I cut the tenon just a fraction smaller than the required diameter so that it will fit neatly into the base area. If its diameter is exactly the same as the diameter of the base, the tenon will be too tight.

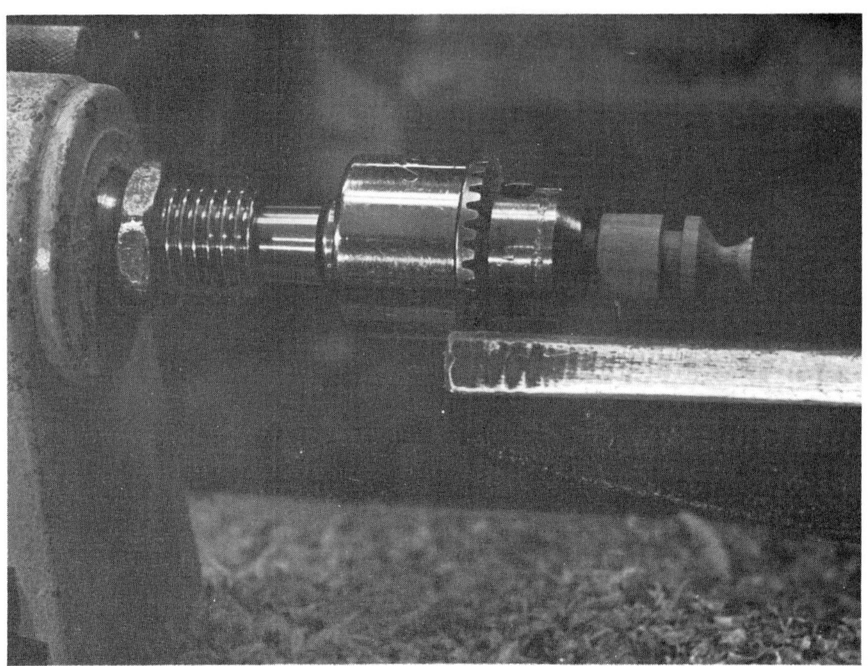

Illus.131. Turned base and lid in Jacob's chuck

3. Using a parting tool, separate the lid and tenon from the base. To prevent reducing the length of the tenon too much, cut off a portion of the base's top section when you are making the cut. Catch the lid and tenon with your other hand as they fall away (Illus. 131).

So that the storage-area hole will be drilled in the exact center of the base, use the point of a skew chisel and allow it to make a small indentation in the top surface of the base. This indentation will be the entry point for the drill bit.

4. Separate the base from the tenon using a parting tool. Place the parting tool as near the chuck as possible when you are making the cut. This will allow you to have a base that may be slightly longer than you anticipated. Make the cut straight so that the base will stand erect. Then catch the base as it separates from the chuck tenon. You may have to sand some wood strands from the bottom surface of the base.

5. It's best to drill the storage-area hole in the base using a drill press. To keep the base from turning during the drilling procedure, I wear a leather glove and hold the base firmly in place. Adjust the drill press to drill the hole, leaving a bottom thickness of at least ⅛". If you use a brad-point bit, be certain the point doesn't penetrate the bottom surface. With the hole drilled, test the lid and tenon for fit.

For a discussion of finishing products and procedures, refer to Chapter 43.

42
TINY DECORATIVE TURNINGS

When I don't have the time or energy to turn a large project, spindle-turning tiny pieces can be an extremely relaxing and enjoyable experience. Some of these tiny turnings have a function—for example, the birthday cake candle-holder in Illus. 132; but others are simply interesting little pieces that can be used to decorate a shelf.

One of the satisfying aspects of spindle turning is that you don't always have to make a

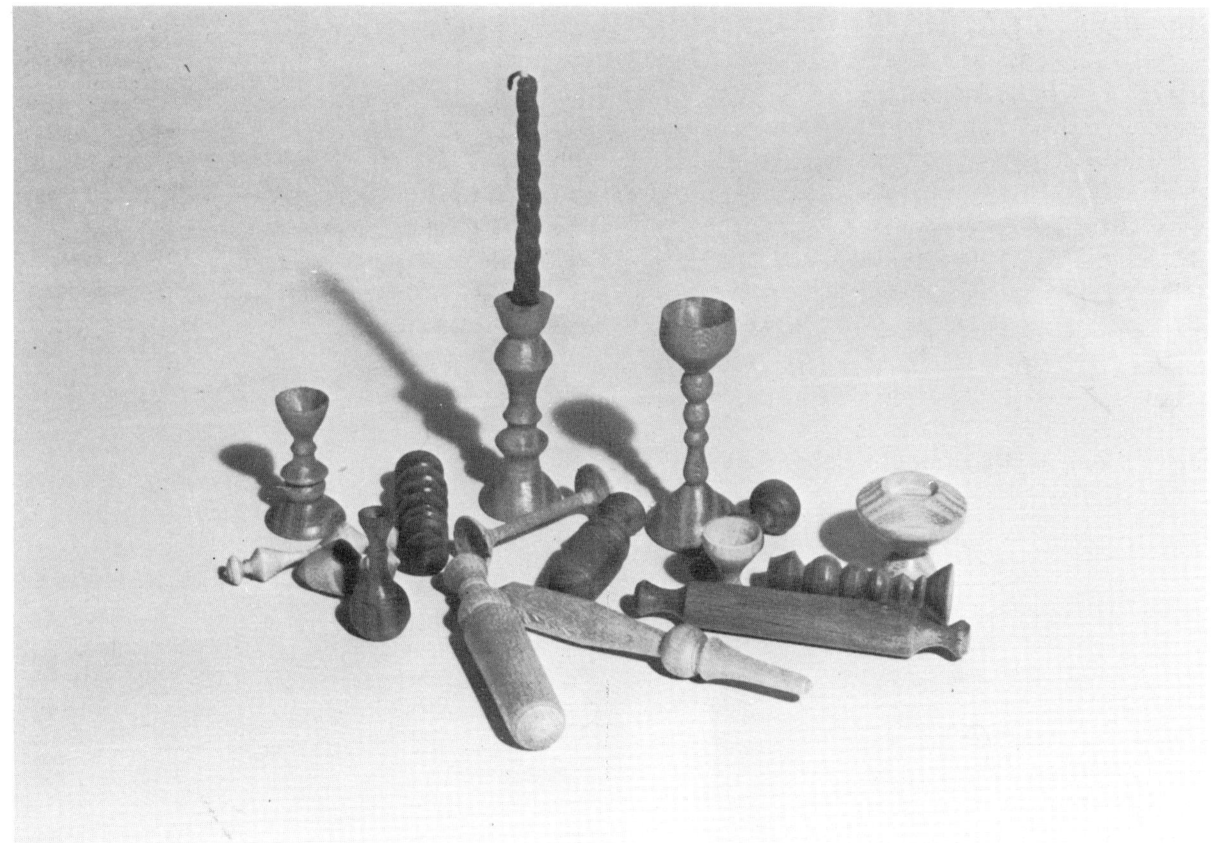

Illus.132. Tiny decorative turnings

Illus.133. Tiny piece in Jacob's chuck

recognizable object. You can place a spindle between centers or one in a chuck and simply make beads or coves, and you will still get the satisfaction that comes with turning wood.

With a ½″ Jacob's chuck mounted on the lathe, cut a quantity of ½″-diameter dowels to a 2″ to 4″ length. Secure a dowel in the chuck; then choose a tool and do some free-form spindle-turning. I think you will be amazed at how much fun you can have with a few small pieces of dowel. After you've turned a few dowels, undoubtedly a design or an object will come to mind. Once this takes place, you're well on your way towards turning a tiny decorative piece from your own imagination (Illus. 133).

43
FINISHING

The chemical industry has made great strides in producing an array of quality finishing products. But the very quantity of available products has actually made the finishing process more complex and confusing for many woodturners. To minimize my own frustration, I use only two products and a few simple procedures.

While I generally prefer off-lathe finishing, some pieces can be finished on the lathe. If you want to use a wax or shellac finish, on-lathe finishing is probably the best. This method allows you to properly apply the finishing product and, when you apply a rag to the surface, the speed of the lathe creates a bright shiny finish. You can also effectively apply some oil finishes using the on-lathe method.

PRODUCTS

Finishing products generally fall into two categories: those that are absorbed into the wood and those that remain on the surface. You get very different results with these different types of finishes.

The finishing products that are absorbed into the wood are generally the various oil finishes. These finishes give a piece a very warm, natural appearance. It's easy to apply and maintain oil finishes. And some of them are nontoxic; so you can apply them to pieces that will be used for food. Whenever I want an oil finish on a piece, I use Watco Danish oil. I've had excellent results with this product and it's very easy to use.

The finishes that remain and build on the surface of a piece are generally the various lacquers and polyurethanes. You will find that a large variety of these products is on the market. However, I tend to use Deft, which is a high-quality lacquer, because it dries quickly and leaves a semigloss finish. Deft is ideal if you finish your turnings in your shop. Since it dries so quickly, it won't attract and hold the dust and debris that are always present in woodworking shops. I usually apply it with a brush, but Deft is also available in spray cans.

In addition to Watco Danish oil and Deft, I use Trewax or Butcher's paste wax on most of my turnings. You want to use the clear wax that both manufacturers have available. Even though it's a surface finish, wax greatly enhances the overall appearance of a piece that you have finished with Watco Danish oil or Deft.

PROCEDURES

Before you apply any finishing product, you need to read the directions on the label. *Most finishing products should be used in a well-ventilated area and the rags used with them properly disposed of, since they are very prone to spontaneous combustion.* If you plan to be exposed for an extended period of time to any type of finishing product, you should wear a breathing device that will filter out the dangerous fumes. If you plan to finish pieces in your basement, be certain you have adequate ventilation. Pilot lights on gas furnaces and hot-water heaters can present a potential danger if you are using finishing products in a closed basement area. If at all possible, finish your turnings outside or in a well-ventilated garage.

Whenever a piece will be used with food, always apply a nontoxic finish, such as Watco Danish oil. For example, use Watco on the icebreaker project presented in Chapter 8. You may also want to use Watco on the soap-dish project presented in Chapter 33, because Watco fills wood pores, making a piece somewhat water- and chemical-resistant. Watco is also a good choice for the various desk-top projects and the different-sized bowls.

When you are using an oil finish, the piece needs to be very well sanded. If the surface is not properly prepared, small wood hairs tend to be exposed and this will give the piece a very dull, mousey appearance. Rubbing the surface with 600-grit wet-or-dry abrasive paper helps to eliminate some of these hairs. I have found that a piece that will be oiled requires a better-prepared surface than a piece that will be covered with a surface finish.

The various large boxes, especially if they have been turned from black walnut, lend themselves to an oil finish. I generally use a Watco finish on the bean-pot box design presented in the first chapter. This project seems to be frequently handled and rubbed, and oil finishes are ideal for pieces that will be handled a great deal because the handling tends to enhance the oil finish.

A distinct advantage of using an oil finish is that, if a piece gets scratched, the scratch can easily be removed. When you use a surface finish, a scratch will remove a portion of the actual finish; this doesn't happen with an oil finish. All you need to do is apply some additional oil finish and rub it in a bit, and the scratch will disappear. You can also revitalize and restore a piece you've made some time ago to its original appearance by applying a few new applications.

I usually do not use Watco on the various spalted woods. It's almost impossible to obtain a quality finish with an oil on the spalts, especially on the more porous woods. However, you may want to consider using an oil finish on those pieces of spalted wood that remain very hard despite the activity of the fungi. For example, spalted hard maple frequently has brilliant black zone lines, but also remains very hard. But more often than not, the spalts are too porous and will simply absorb an unceasing amount of oil. The end grains especially tend to be a problem when it comes to soaking up the oil.

The various projects made from pine, when properly sanded, are greatly enhanced by an oil finish. For example, the large turned plates, with or without the inlay, look especially beautiful with an oil finish.

In addition to the projects I've mentioned, you will find that many other pieces look superb with an oil finish. If you've never used Watco Danish oil (or a similar oil finish), obtain the smallest possible quantity and experiment with it. Following the directions on the label, apply the oil to some scraps of different kinds of wood and see what kind of finish develops.

For surface finishing, I generally apply Deft in a rather haphazard fashion. Since it's such a quick-drying lacquer, Deft is easy to apply off the lathe. For this high-quality finish, it's best to use the more expensive natural bristle brushes, although I tend to use some of the least expensive brushes myself. Needless to say, with the inexpensive brushes, you must be prepared to occasionally pick off a few brush hairs from the piece you're finishing. On the other hand, unlike expensive brushes that need to be cleaned and saved, my brushes can be thrown away when the finishing is completed. Before you begin to finish a piece with Deft, be sure to read the directions on the label. Proper ventilation is imperative when you are using this product.

Whereas I brush Deft on most of my turnings, Andy Matoesian likes to use the spray cans instead. When he is finishing his turned pens, for example, he simply places them on a finishing jig and sprays them with a can of Deft. The finishing jig simply consists of nails driven through a piece of scrap material. Andy places the pens, with their insert holes, over the nails, which hold them upright for the finishing process. When you use Deft in a spray can, be sure to read and follow the directions. It's very easy to overspray or hold the can too near to the piece. *You should also wear a breathing device whenever you are spraying a finishing product.*

When I finish a turning with Deft, I usually do so in stages. For example, with a box, I first apply Deft to the inside and outside of the base. Then I place the base on a piece of scrap material until the Deft is dry. While the base is drying, I apply Deft to the top surface of the lid and then place the lid on a piece of scrap until it's dry. I apply three coats, using this procedure. When the third coat is sufficiently dry, I rub the finished surfaces with (0000) steel wool. Using an air compressor and gun or a clean rag, I then blow or wipe away the finishing dust and the steel hairs.

As a rule, I apply two additional coats of Deft to the base-and-lid areas. However, you should be certain the rubbed surfaces are clean before you begin applying more Deft. After the fifth and final coat, I again rub the surfaces with (0000) steel wool. You should rub steel wool with the grain if possible. Do not apply too much pressure with the steel wool or you will cut through the various layers of finish. I use the steel wool because it removes the small particles that get attached to the finish. The steel wool also tends to dull the glossiness of the finish.

Using these same procedures, apply Deft to the bottom surfaces of the base and lid. Be careful not to apply too much Deft or it will run onto the surfaces you have already finished. If this becomes a problem, do not use the steel wool for the final applications on the other surfaces until you have applied Deft to all the surfaces. You also need to handle the pieces very carefully so that you don't scratch or mar the finished surfaces.

After I have applied Deft to the base and lid and rubbed the surfaces with steel wool, I apply either a clear Trewax or Butcher's paste wax. Apply a thin coating of the wax to all surfaces and then buff or polish when the wax is partially dry. As a rule, I apply two thin coats of paste wax and polish vigorously with a flannel cloth. The wax especially helps protect the surface finish from water spots.

Projects that require many additional coats of Deft for an acceptable finish are those made from spalted woods or those with textured surfaces. It's frequently very difficult to build up a surface finish on spalted woods using Deft. Since Deft is so thin, it gets quickly absorbed into the porous wood. Very often I will apply as many as 10 coats of Deft to a spalted bowl, along with a few additional coats to the cross-grain areas. Finishing the spalts can take a great deal of time and can use up quantities of Deft, but it's well worth the effort. The same holds true with textured wood. With the textured bowl presented in Chapter 5, you need to keep applying the finish until you achieve an acceptable surface.

Surface finishes, such as Deft, can cause some woods to bleed on others. For example, when I apply a lacquer to padouk, it will turn red because of the sawdust in the wood. If I use too much lacquer, the red lacquer will run onto and stain any surrounding wood. Therefore, when I use an inlay in a project, especially if it's made from padouk, I will apply two light coats of Deft to it first; this tends to seal the wood and prevent any future problems. If one wood does bleed onto another, you need to sand off the finish and start over again.

When you are working on projects that will have items secured to them, you must apply Deft before the items are secured in place. For example, you should finish the desk-top projects that employ pens and funnels before you screw these items in place. If you attach these kinds of items to a project prior to finishing, you will inevitably get a certain amount of finish on them and this will make your project look sloppy.

While the foregoing material may seem somewhat simplistic given the plethora of finishing products on the market and the sophistication of some finishing techniques, these products and procedures have served me well. They are certainly adequate for finishing the range of small turned projects presented in this book. Although I continue to try new products and methods, I find that I inevitably return to using Watco Danish oil and Deft and the various methods described here.

METRIC EQUIVALENCY CHART

mm—millimetres *cm—centimetres*

INCHES TO MILLIMETRES AND CENTIMETRES

inches	mm	cm	inches	cm	inches	cm
1/8	3	0.3	9	22.9	30	76.2
1/4	6	0.6	10	25.4	31	78.7
3/8	10	1.0	11	27.9	32	81.3
1/2	13	1.3	12	30.5	33	83.8
5/8	16	1.6	13	33.0	34	86.4
3/4	19	1.9	14	35.6	35	88.9
7/8	22	2.2	15	38.1	36	91.4
1	25	2.5	16	40.6	37	94.0
1 1/4	32	3.2	17	43.2	38	96.5
1 1/2	38	3.8	18	45.7	39	99.1
1 3/4	44	4.4	19	48.3	40	101.6
2	51	5.1	20	50.8	41	104.1
2 1/2	64	6.4	21	53.3	42	106.7
3	76	7.6	22	55.9	43	109.2
3 1/2	89	8.9	23	58.4	44	111.8
4	102	10.2	24	61.0	45	114.3
4 1/2	114	11.4	25	63.5	46	116.8
5	127	12.7	26	66.0	47	119.4
6	152	15.2	27	68.6	48	121.9
7	178	17.8	28	71.1	49	124.5
8	203	20.3	29	73.7	50	127.0

INDEX